Research Infrastructures for Hardware Accelerators

Synthesis Lectures on Computer Architecture

Editor
Margaret Martonosi, *Princeton University*

Synthesis Lectures on Computer Architecture publishes 50- to 100-page publications on topics pertaining to the science and art of designing, analyzing, selecting and interconnecting hardware components to create computers that meet functional, performance and cost goals. The scope will largely follow the purview of premier computer architecture conferences, such as ISCA, HPCA, MICRO, and ASPLOS.

Research Infrastructures for Hardware Accelerators
Yakun Sophia Shao and David Brooks
2015

Analyzing Analytics
Rajesh Bordawekar, Bob Blainey, and Ruchir Puri
2015

Customizable Computing
Yu-Ting Chen, Jason Cong, Michael Gill, Glenn Reinman, and Bingjun Xiao
2015

Die-stacking Architecture
Yuan Xie and Jishen Zhao
2015

Single-Instruction Multiple-Data Execution
Christopher J. Hughes and
2015

Power-Efficient Computer Architectures: Recent Advances
Magnus Själander, Margaret Martonosi, and Stefanos Kaxiras
2014

FPGA-Accelerated Simulation of Computer Systems
Hari Angepat, Derek Chiou, Eric S. Chung, and James C. Hoe
2014

Multi-Core Cache Hierarchies
Rajeev Balasubramonian, Norman P. Jouppi, and Naveen Muralimanohar
2011

A Primer on Memory Consistency and Cache Coherence
Daniel J. Sorin, Mark D. Hill, and David A. Wood
2011

Dynamic Binary Modification: Tools, Techniques, and Applications
Kim Hazelwood
2011

Quantum Computing for Computer Architects, Second Edition
Tzvetan S. Metodi, Arvin I. Faruque, and Frederic T. Chong
2011

High Performance Datacenter Networks: Architectures, Algorithms, and Opportunities
Dennis Abts and John Kim
2011

Processor Microarchitecture: An Implementation Perspective
Antonio González, Fernando Latorre, and Grigorios Magklis
2010

Transactional Memory, 2nd edition
Tim Harris, James Larus, and Ravi Rajwar
2010

Computer Architecture Performance Evaluation Methods
Lieven Eeckhout
2010

Introduction to Reconfigurable Supercomputing
Marco Lanzagorta, Stephen Bique, and Robert Rosenberg
2009

On-Chip Networks
Natalie Enright Jerger and Li-Shiuan Peh
2009

The Memory System: You Can't Avoid It, You Can't Ignore It, You Can't Fake It
Bruce Jacob
2009

Fault Tolerant Computer Architecture
Daniel J. Sorin
2009

The Datacenter as a Computer: An Introduction to the Design of Warehouse-Scale Machines
Luiz André Barroso and Urs Hölzle
2009

Computer Architecture Techniques for Power-Efficiency
Stefanos Kaxiras and Margaret Martonosi
2008

Chip Multiprocessor Architecture: Techniques to Improve Throughput and Latency
Kunle Olukotun, Lance Hammond, and James Laudon
2007

Transactional Memory
James R. Larus and Ravi Rajwar
2006

Quantum Computing for Computer Architects
Tzvetan S. Metodi and Frederic T. Chong
2006

Research Infrastructures for Hardware Accelerators
Yakun Sophia Shao and David Brooks

ISBN: 978-3-031-00622-7   paperback
ISBN: 978-3-031-01750-6   ebook

DOI 10.1007/978-3-031-01750-6

A Publication in the Springer series
SYNTHESIS LECTURES ON COMPUTER ARCHITECTURE

Lecture #34
Series Editor: Margaret Martonosi, *Princeton University*
Series ISSN
Print 1935-3235   Electronic 1935-3243

Research Infrastructures for Hardware Accelerators

Yakun Sophia Shao and David Brooks
Harvard University

SYNTHESIS LECTURES ON COMPUTER ARCHITECTURE #34

ABSTRACT

Hardware acceleration in the form of customized datapath and control circuitry tuned to specific applications has gained popularity for its promise to utilize transistors more efficiently. Historically, the computer architecture community has focused on general-purpose processors, and extensive research infrastructure has been developed to support research efforts in this domain. Envisioning future computing systems with a diverse set of general-purpose cores and accelerators, computer architects must add accelerator-related research infrastructures to their toolboxes to explore future heterogeneous systems. This book serves as a primer for the field, as an overview of the vast literature on accelerator architectures and their design flows, and as a resource guidebook for researchers working in related areas.

KEYWORDS

accelerators, specialized architecture, SoC, high-level synthesis, simulators, design space exploration, workload characterization, benchmarks

Contents

Preface

Specialized architectures have been a growing topic in both academic research and commercial development for the past decade. As traditional technology scaling slows, specialization becomes a viable solution for computer architects to continue performance growth and energy efficiency improvements without relying on technological advances.

This book aims to present a high-level overview of the state-of-the-art accelerator research in both industry and academia, with a special emphasis on research infrastructure available for accelerator-related research. This book begins by describing the technology trends that have led accelerator research to prominence. In Chapter 2, we present a taxonomy of accelerator research and practice, with the goal of introducing the reader to the flavor of accelerator designs that have been proposed in recent years. Chapter 3 presents the standard accelerator design flow from RTL generation, simulation, and synthesis. Recent advances in high-level synthesis (HLS) tools provide a promising path for accelerator development in the future, and we describe the capabilities of commercial tools like Xilinx's Vivado HLS and their limitations. Chapter 4 discusses pre-RTL modeling approaches to facilitate the rapid exploration of the design space of accelerators as well as the interaction between accelerators and the rest of the system. Chapter 5 focuses on workload characterization approaches in the context of accelerators and Chapter 6 discusses benchmarking. We end this book with a discussion on the challenges and opportunities of accelerator architectures and design tools in Chapter 7.

Yakun Sophia Shao and David Brooks
October 2015

Acknowledgments

We would like to thank Margaret Martonosi for encouraging us to write this book, as well as the feedback and support she has provided throughout the project. We would also like to thank Michael Morgan for providing us this opportunity and keeping us on schedule during the whole process. Many thanks to Kelly Shaw, Luis Ceze, and Glenn Holloway for their detailed comments that were invaluable in improving this manuscript. We would especially like to thank our many collaborators over the years: Gu-Yeon Wei, Viji Srinivasan, Simone Campanoni, Michael Lyons, Brandon Reagen, Sam Xi, and Robert Adolf. Much of the content of the book is built on wonderful collaborations and insightful discussions with many of them.

Yakun Sophia Shao and David Brooks
October 2015

CHAPTER 1

Why Accelerators, Now?

"It was the best of times, it was the worst of times, it was the age of wisdom, it was the age of foolishness, it was the epoch of belief, it was the epoch of incredulity, it was the season of Light, it was the season of Darkness, it was the spring of hope, it was the winter of despair." Dickens, 1859.

The era of heterogeneous computing is upon us. Heterogeneity comes in many forms including domain-specific processors and application-specific accelerators. Almost all major semiconductor vendors have chips that include accelerators, big or small, for a variety of applications. Research in this space has grown in popularity, and there is a vibrant community of researchers in the fields of computer architecture and VLSI-CAD that seek to embrace this new era. The community is ushering in new architectures, tools, and evaluation methodologies, and this Synthesis Lecture expounds upon these developments. But what exactly is an accelerator and why do we need accelerators now? In the following sections, we seek to define accelerators and discuss how fundamental changes in the semiconductor industry have brought about the emergence of this new era of hardware accelerators.

1.1 WHAT IS AN ACCELERATOR?

An accelerator is a specialized hardware unit that performs a set of tasks with higher performance or better energy efficiency than a general-purpose CPU. Examples of accelerators include digital signal processors (DSPs), graphics processing units (GPUs) and fixed-function application-specific integrated circuits (ASICs) like video decoders. The use of accelerators is not a new idea; the deployment of floating point co-processors in the 1980s marked one of the early adoptions of accelerators, and accelerators have been a common feature of system-on-chip (SoC) architectures for embedded systems for decades. However, accelerator research and deployment in mainstream computer architectures has not been embraced until very recently. This begs the question: Why accelerators, now?

To answer this question, let us first look at the interaction between computer architecture and semiconductor technology. Computer architecture sits as a layer between the semiconductor fabrication and circuit technology used to produce chips and the high-level system software and applications that run on them. Hence, every major architectural breakthrough is deeply rooted in the advance of underlying technology and the quest for better performance or energy efficiency for

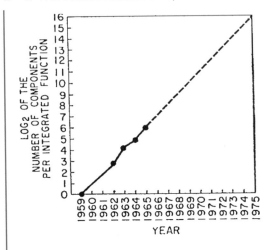

Figure 1.1: Moore's law [89].

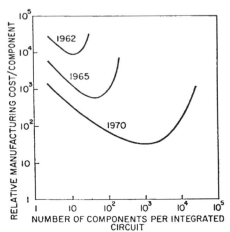

Figure 1.2: Moore's cost data [89].

applications. In the rest of this chapter, we discuss the historical scaling trends that have driven the growth of the semiconductor industry and how disruption in these trends has led to the growing popularity of accelerators.

1.2   A TALE OF TWO SCALINGS

The semiconductor industry has been driven by two scaling laws: Moore's Law and Dennard scaling. It is these two scaling trends that have resulted in the popularity of CMOS technology and subsequent advances in computing technology over the past several decades.

1.2.1   MOORE SCALING

The semiconductor industry has recorded impressive achievements since 1965, when Gordon Moore published the observation that would become the industry's guiding standard for the subsequent five decades [89]. Moore's law states that the number of transistors that can economically be fit onto an integrated circuit doubles every two years, demonstrated in the well-known plot shown in Figure 1.1. This simple plot set the pace for progress in the semiconductor industry, although Moore's scaling prediction is purely an empirical observation of technological progress.

Moore's law is about more than just shrinking transistor sizes and better integration capability. It is fundamentally a cost-scaling law. Moore derived the law as he was ultimately interested in shrinking transistor costs. Figure 1.2 is also from Moore's original paper, and is an often overlooked observation that is every bit as important as transistor density trends:

"For simple circuits, the cost per component is nearly inversely proportional to the number of components, the result of the equivalent piece of semiconductor in the equivalent package containing more components. But as components are added, decreased yields more than compensate for the increased complexity, tending to raise the cost per component. Thus there is a minimum cost at any given time in the evolution of the technology."

What Moore observed is that the cost of transistors depends on two factors. One is the density of transistors that we can cram onto a single chip, and the other one is the yield rate of the fabrication. Increasing the number of transistors per silicon area makes transistors cheaper. However, increasing the amount of silicon area dedicated to an individual chip also increases the chances of a chip being rendered inoperable by defects found on the wafers, which in turn will drive down the yield rate.

To maintain Moore's cost scaling, two factors are critical:

1. transistor size—the smaller the better since we can cram more on the same area; and

2. wafer size—the larger the better since we can produce more chips from a fixed number of processing steps. Also, empirically defects are more likely to occur at the edge of a wafer, so a larger wafer also means smaller defective densities.

Over the years, Moore's law has driven the semiconductor industry to achieve higher integration density and lower the cost of the transistor by scaling the sizes of transistors and wafers together with a combination of other innovative "circuit and device cleverness" to increase the yield [104].

1.2.2   DENNARD SCALING

Moore's analysis of scaling only stated that we need to make transistors smaller but it did not touch on how to make this happen. It is Dennard scaling that outlined how we make smaller transistors at each technology generation. Robert Dennard addressed this in his paper on metal-oxide semiconductor (MOS) device scaling in 1974 [49]. In this paper, Dennard showed that when voltages are scaled along with transistor dimensions, a device's electric fields remain constant, and most device characteristics are preserved. Table 1.1 summarizes transistor and circuit parameter changes under ideal scaling conditions, where k is a unit-less scaling constant. In a new technology generation, the transistor dimensions, e.g., gate oxide thickness (t_{ox}), length (L), and width (W), become smaller compared to the previous generation by a factor of $1/k$. As transistors get smaller, they switch faster, use less power but achieve the same power density. Dennard's scaling has set the roadmap for the semiconductor industry for each generation of process technology, with a concrete transistor scaling formula to move each generation forward.

Table 1.1: Dennard scaling rules

Device or Circuit Parameter	Scaling Factor
Device dimension t_{ox}, L, W	$1/k$
Doping concentration N_a	k
Voltage V	$1/k$
Current I	$1/k$
Capacitance eA/t	$1/k$
Delay time per circuit VC/I	$1/k$
Power dissipation per circuit VI	$1/k^2$
Power density VI/A	1

1.3   THE COMBINATION OF MOORE AND DENNARD SCALING

As we discussed earlier, Moore's scaling fundamentally is about cost scaling: the more transistors that can be packed into a given silicon area, the cheaper it is to fabricate a transistor. On the other hand, Dennard's scaling is a performance scaling law. It described how transistors have better delay and power characteristics as they get smaller. Every scaling theory faces two possibilities: to continue or to end. By combining the two possibilities with Moore's and Dennard's scaling theories, we end up with a combination of four scaling trends, shown in Figure 1.3.

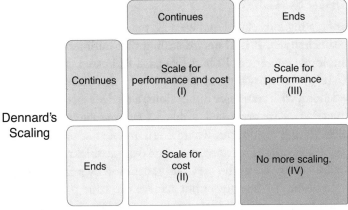

Figure 1.3: Four relations of Moore and Dennard scaling.

1. Both Moore and Dennard scalings continue—Scale for both performance and cost (Region I). This is the ideal scaling region, which the semiconductor industry enjoyed up until the 2000s, resulting in both faster and cheaper transistors.

2. Moore scaling only—Scale for cost (Region II). The state of the semiconductor industry since the mid 2000s: Dennard scaling stops but we still have cheaper transistors per generation.

3. Dennard scaling only—Scale for performance (Region III). This scaling region has not been realized in an economically practical way.

4. No more scaling—Scale for nothing (Region IV). CMOS technology will become a commodity, likely resulting in lower profits for fabrication companies. In this case, there will be no motivation to scale until a new transistor technology emerges to displace end-of-CMOS devices.

We will discuss each of the four trends and how they have shaped the evolution of the computing industry.

1.3.1   MOORE + DENNARD—WHERE WE WERE

The semiconductor industry reaped the benefits from scaling in Region I for a long while, where both Moore and Dennard scaling provided nearly ideal benefits to chip designers. Figure 1.4 shows that minimum feature sizes scaled consistently over a 30 year period from the mid 1970s to the mid 2000s, closely tracking with Dennard scaling projections and providing designers with smaller, faster transistors. At the same time, Moore scaling, represented by a doubling of the number of transistors every two years (Figure 1.5), led to an exponential decrease in cost per transistor (Figure 1.6).

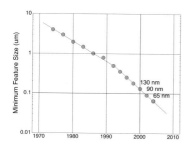

Figure 1.4: Transistor size [31].

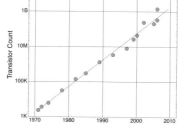

Figure 1.5: Transistor count [31].

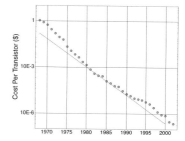

Figure 1.6: Cost per transistor [90].

1.3.2   MOORE SCALING ONLY—WHERE WE ARE

By the middle part of the 2000s, the semiconductor industry realized that scaling trends had created a major problem with power consumption that made it difficult to economically cool microprocessors. The fundamental reason is that while the industry was successful in matching Dennard scaling projections in producing smaller transistors, the chip supply voltage had not kept pace with the theoretical projections.

Figure 1.7 shows historically how supply voltage scales with transistor feature size. We notice that from the 0.13 *um* generation on, supply voltage has slowed down, which is tightly coupled to the lack of threshold voltage scaling in devices to counter sub-threshold leakage current. Thus, we observed an abrupt and drastic close to the clock frequency scaling era in the early 2000s, shown in Figure 1.8. As Moore's law continues to provide increasing number of transistors per generation, the microprocessor industry adopted multicore architectures that use many simpler processors to keep energy per instruction low while increasing the aggregated performance of the entire chip through thread-level parallelism.

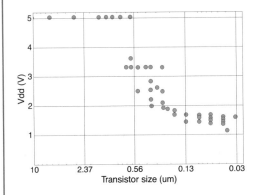

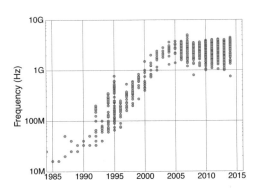

Figure 1.7: Supply voltage stops scaling [48].

Figure 1.8: Frequency stops scaling [48].

However, the multicore approach does not fundamentally extend Dennard's scaling. On the one hand, the overall speedup is still heavily limited by the sequential portion of the application [19]. Figure 1.9 illustrates the ideal speedup with respect to the number of cores for applications with different parallelism. For extreme parallel workloads, e.g., more than 95% portion of the program is parallelizable, the achievable speedup for 256 cores is less than 20. On the other hand, the worsening energy and speed scaling of transistors limits the number of transistors that can be powered on at the same time, leading to transistor under-utilization (i.e., dark silicon [51]). Figure 1.10 illustrates the overall achievable speedup over technology generations. Due to the power and parallelism limitation, a significant gap exists between what is achievable with multi-core scaling and what is expected from Moore's law. The question for architects now is: what's next?

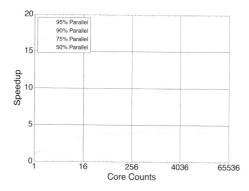

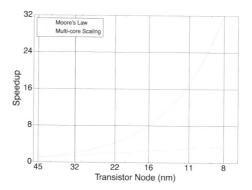

Figure 1.9: Achievable speedup with respect to the parallel portion of the program [19].

Figure 1.10: Multi-core speedup across technology generations compared to Moore's law scaling [51].

1.3.3   DENNARD ONLY—WHERE WE ARE UNLIKELY TO BE

Although it is unlikely that we will enter a regime of Dennard-only scaling, it could exist if there were a new device technology that scaled in performance and energy, but not in the economic dimension of Moore scaling. For the sake of discussion, such a technology would likely be confined to niche applications that need ultra high performance or low energy consumption and are not cost sensitive. Thus, it is likely that only a small subset of semiconductor players would be interested in exploring such technologies, as the capital and R&D expenses of maintaining scaling would be hard to overcome.

1.3.4   A FUTURE WITHOUT SCALING: "THE WINTER OF DESPAIR"

A return of Region I scaling is also unlikely in the medium term given device technology trends and projections over the next decade. In fact, traditional Moore scaling is already measurably slowing down. Figure 1.11 shows Intel's historical technology scaling trend based on the release date of the first microprocessor in each technology node, along with projections for the arrival of the 10nm node. Over the past decade, Intel has been following its famous tick-tock development cycle: first releasing processors with an existing architecture but a new technology node (tick), and then the following year releasing processors with a new architecture on the then-mature technology node (tock). In this way, Intel can release new products every year, alternating architecture and technology improvements. However, the introduction of Intel's 14 nm process in the fourth quarter of 2014 was a disruption of this rhythm as it was delayed for half of a year beyond the original projection of the second quarter. Moreover, in July 2015 Intel announced that the 10 nm node will not be ready until the second half of 2017, which is another significant delay [7]. In this case, there will be three generations of products using the same 14 nm technology.

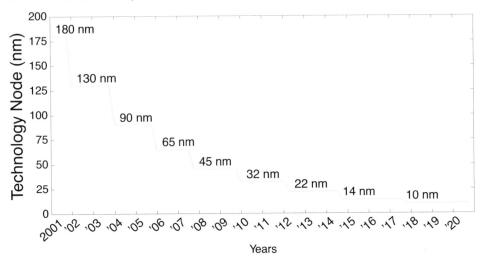

Figure 1.11: Intel historical technology scailing trend and projections [6, 7].

Intel's latest delay is part of a larger trend in semiconductor manufacturing. Switching to a new process node is getting more complex and more expensive than ever. There has been a drastic increase in two categories of fabrication costs. First, the R&D cost for developing next generation CMOS fabs is increasing rapidly. Innovations like FinFETs have allowed the semiconductor industry to continue to scale transistor feature sizes. However, the fabrication process is getting significantly more complex due to techniques such as multi-patterning [117, 129]. Such costs overcome lithography bottlenecks at the expense of more fabrication steps, expensive tools, and higher mask design costs. Several semiconductor companies have become fabless in the past several years due to the increasing cost of maintaining state-of-the-art fabrication facilities, and it is likely that we will see even more consolidation in the industry.

On the other hand, it is also getting harder to increase the wafer size. Larger wafer size can lower fabrication costs by increasing the number of dies per wafer and providing better yields. However, the cost of equipment grows significantly with larger wafer sizes. Migrating from today's 300 mm wafers to full-scale 450 mm fab lines will cost $10 billion to $15 billion, and thus this transition is is not expected to occur until 2020 [10].

Figure 1.12 shows the cost of transistors over different technology generations in recent years. The data, provided by Nvidia, shows that there is a minimal projected cost benefit after migrating past 28 nm technology, as the projected curves have normalized cost to 28 nm. Thus, scaling to smaller feature sizes will no longer provide an economic benefit for fabless IC companies. If this comes to pass, it will effectively mean the end of the economic scaling projection of Moore's law.

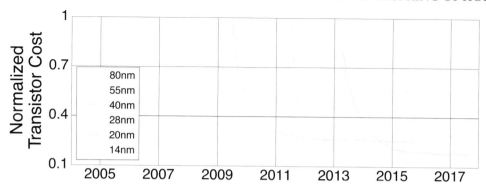

Figure 1.12: Cost per transistor stops scaling [2].

1.4   TO LIVE WITHOUT SCALING: "A SPRING OF HOPE"

The computing industry has begun to adjust to the loss of Dennard scaling, and the loss of Moore scaling is on the near-term horizon. This will likely lead to additional consolidation in the semiconductor industry, and fabrication companies will rely on "More-than-Moore" technologies to provide differentiation [20]. One possible outcome of this is that transistors will become a commodity with potentially lower profits for fabrication companies. Without either kind of scaling, there is also risk of stagnation in the overall computing industry.

At the same time, technology disruptions often mean new opportunities, and in this case, there is significant opportunity for innovation at the design and architecture level. As the computing industry necessarily becomes aware of this disruption, opportunities to push new software development paradigms will abound. Companies will increasingly differentiate their products based on vertically integrated solutions that leverage new applications mapped to innovative hardware architectures. In this context, application- and domain-specific hardware accelerators are one of the most promising solutions for improving computing performance and energy efficiency in a future with little benefit from device technology innovation.

1.4.1   WHY NOT ARCHITECTURAL SCALING?

Even with both Moore and Dennard scaling, architecture-level innovations have already contributed significantly to the performance increase of computing systems. Figure 1.13 demonstrates the normalized performance of multiple generations of processors since Intel's 80386 microprocessor. The blue circles are normalized overall performance; the orange diamonds are normalized FO4 delay, a rough metric for Dennard's transistor performance scaling. We see that roughly half of the performance improvement is due to faster transistors; the other half is mostly from architectural innovations. With the increasing number of transistors on chip, computer architects have been working hard to make use of these transistors for better performance. Examples include superscalar and out-of-order scheduling to increase instruction-level parallelism, better cache hierarchies to overcome memory bottlenecks, multi-core architecture to harness thread-level parallelism, and specialized units, like SIMD, to increase the performance of applications with data-level parallelism.

Still, it is likely that we will not see these same increases in performance and efficiency for general-purpose cores going forward. Many of the above techniques incur unwanted power overheads. Furthermore, with the lack of device scaling, architectural mechanisms will need to provide far higher benefits to provide the differentiation of performance and energy needed to drive the computing industry. The blue squares in Figure 1.13 show how processor performance scaled over time, while the orange diamonds indicate how much speedup came from improving the manufacturing process. Here all the performance is normalized to the performance of the Intel 386. To estimate the performance of a processor if it were manufactured using a newer technology without microarchitecture changes, the authors use the delay of an inverter driving four equivalent inverters (a fanout of four, or FO4) to quantify the speedup achieved through frequency scaling. Figure 1.13 suggests that to overcome the loss of FO4 scaling over the next decade, computer architects will need to deliver an additional order of magnitude performance improvement for the computing industry to enjoy the same amount of performance differentiation that we achieved over the past decade.

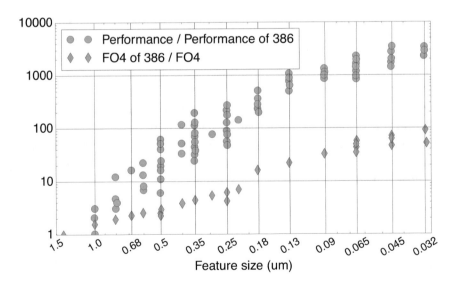

Figure 1.13: Performance increase is more than technology scaling [48].

1.4.2   SPECIALIZATION MAKES A DIFFERENCE

Hardware acceleration in the form of datapath and control circuitry customized to particular algorithms or applications has surfaced as a promising approach, as it delivers orders of magnitude performance and energy benefits compared to general-purpose solutions. Figure 1.14 shows the energy efficiency comparison between general-purpose processors, DSPs, and dedicated application-specific accelerators [130]. The data was collected from 20 different chips across different architectures. These chips were originally published at the International Solid State Circuits Conference (ISSCC) between 1998 and 2002. Compared to general-purpose processors, customized processors like DSPs deliver from 10× to 100× more energy efficiency, while dedicated application-specific accelerators are 1000× more energy efficient.

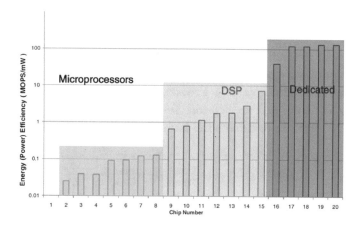

Figure 1.14: Specialization makes a difference.

Customized architectures composed of CPUs, GPUs, and accelerators are already seen in mobile systems and are beginning to emerge in servers and desktops. Analysis of die photos from three generations of Apple's SoCs: A6 (iPhone 5), A7 (iPhone 5S), and A8 (iPhone 6), shows that more than half of the die area is dedicated to non-CPU, non-GPU blocks, shown as Others in Figure 1.15 (left). Most of these blocks are application-specific hardware accelerators. The number of specialized blocks has also increased consistently across generations of Apple's SoCs, as illustrated in Figure 1.15 (right).

1.4.3   A CALL FOR TOOLS IN THE ERA OF ACCELERATORS

The natural evolution of the increasing number of accelerators will lead to a growing volume and diversity of customized accelerators in future systems (Figure 1.16), where a comprehensive assessment of potential benefits and trade-offs across the entire system will be critical for system designers.

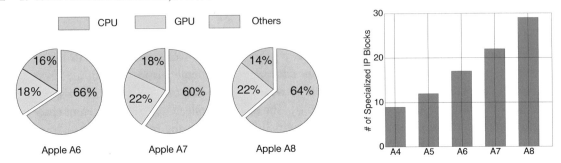

Figure 1.15: Die photo analysis across generations of Apple's SoCs. Left: Die area breakdown. Right: Number of specialized IP blocks.

Historically, the computer architecture community has focused on general-purpose processors, and extensive research infrastructure has been developed to support research efforts in this domain, such as power-performance modeling [24, 29, 32, 79], benchmarking [112, 118], and workload characterization [50, 64]. Envisioning future computing systems with a diverse set of general-purpose cores and accelerators, researchers must add research infrastructures for accelerators to their toolboxes to explore the future heterogeneous, accelerator-centric systems.

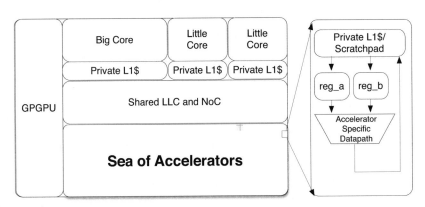

Figure 1.16: Future Heterogeneous Architecture.

> The focus of this book is to describe the state-of-the-art research infrastructures available for accelerator research, including workload analysis tools, architectural models and simulators, and hardware design environments.

CHAPTER 2

A Taxonomy of Accelerators.

"A new scientific truth does not triumph by convincing its opponents and making them see the light, but rather because its opponents eventually die and a new generation grows up that is familiar with it." Max Planck

2.1 NOT ALL APPLES ARE ALIKE

There is increasing interest in accelerator architectures, both in academia and industry. This is an emerging area; different groups or companies have different views of accelerators. In this section, we propose a taxonomy of accelerators with examples from state-of-the-art accelerator research and commercial products.

2.2 ACCELERATOR TAXONOMY

We characterize the space of accelerator designs with two dimensions: *coupling* and *granularity*.

Coupling   defines where accelerators are deployed in the system. Today's system hierarchy typically includes a pipelined processor core with multiple levels of caches attached to the memory bus and then connected to I/O devices through the I/O bus. Conceptually, accelerators can be attached to all levels of this hierarchy, though with unique challenges and opportunities: tightly coupled accelerators require more modifications of the host processor designs, but promise lower invocation latency, while loosely coupled accelerators, often incurring high invocation cost, are freer from the design constraints of the host core. Here we discretize the degree of coupling into the following categories:

1. accelerators that are part of the pipeline;

2. accelerators that are attached to cache;

3. accelerators that are attached to the memory bus; and

4. accelerators that are attached to the I/O bus.

Granularity   defines what kinds of computation are offloaded to accelerators. Finer-grained accelerators are more likely to be used by a variety of applications, but they require certain changes

Table 2.1: Accelerator taxonomy

	Part of the Pipeline	Attached to Cache	Attached to the Memory Bus	Attached to the I/O Bus
Instruction-Level	**FPU, SIMD,** DySER [58],	Hwacha [76, 95, 121], CHARM [43, 44],		
Kernel-Level	NPU [52], 10x10 [40], Convolution Engine [98], H.264 [61],	SNNAP [91], C-Cores [119],	Database [35], Q100 [126], LINQits [41], AccStore [86],	
Application-Level	**x86 AES [18], Oracle/Cavium Crypto Acc [11, 69],**	Key-Value Stores [87], Memcached [80],	Sonic3D [103], DianNao [37, 38], HARP [125], **TI OMAP5 [16], IBM PowerEN [71], IBM POWER7+ [30],**	**GPU, Catapult [97], IBM Power8 CAPI Acc [4],**

to the software stack (i.e., ISAs, OSes, compilers, or programming languages), to allow applications to be decomposed into fine-grained regions that can be implemented by hardware accelerators when appropriate. On the other hand, coarser-grained accelerators are intended to accelerate specific functions or kernels where achieving high efficiency supersedes the need to accommodate more programs. As chips acquire more transistors than can be powered simultaneously [51], using some of those transistors to speed up an application that really matters becomes more affordable. Here we break down the computation granularity into three categories:

1. instruction-level accelerators designed for single primitives like arithmetic operators (including sqrt, sin/cos);

2. kernel-level accelerators that compose key parts of important applications. Examples include matrix multiply, stencil, and FFT; and

3. application-level accelerators that execute the entire applications such as H.264 video decoding and deep neural networks (DNN).

The Taxonomy

Table 2.1 presents the taxonomy of the state-of-the-art accelerator architecture space. Accelerators in bold are industry products, and the rest are research prototypes. Interestingly, we notice that industry products tend to reside at the top left and the bottom right corners of the table (the only exception being cryptography accelerators at the bottom left) while research projects are sprinkled almost everywhere. The reason behind this trend lies in the cost of integration.

Industry has been fairly successful in providing loosely coupled, application-level accelerators, like GPUs or FPGA accelerators connected to the PCIe bus. Such accelerators provide an off-the-shelf solution that minimally interferes with the hardware design of general-purpose cores or the existing CPU software stack. On the hardware side, today's commercial accelerators require minimal changes in general-purpose core designs; accelerators either reside on a separate chip, as in the case of GPUs or FPGA accelerators, or they are plugged in as standalone IP blocks, as in the case of today's SoCs. On the software side, the software stack is also more or less intact: the entire workload is offloaded to accelerators, and no additional management is needed.

Industry has also been good at providing tightly coupled, instruction-level accelerators, like FPU and SIMD units, a category located at the opposite corner of the table from loosely coupled, application-level accelerators. The wide adoption of accelerators in this category also stems from the relative ease with which they can be integrated. Of course, certain modifications to ISAs and compilers are needed to allow applications to leverage accelerators inside the pipeline, such as auto vectorization for vector units, and hardware designers need to balance the new execution units with the rest of the pipeline. However, the changes are well contained inside the pipeline, with little impact on the memory system.

The accelerator space that requires more attention in the next decade lies in cache-attached, kernel-level accelerators. These are the areas where we see little industry presence yet and mostly early work from research projects. In terms of coupling, tightly coupled cache-attached accelerator frees programmers from worrying about low-level data movement between accelerators and cores, as the case of loosely coupled accelerators with DMA, but it needs to cope with non-uniform memory latency, virtual memory, and cache coherence on their own. In terms of granularity, kernel-level accelerators attempt to strike a balance between speedups and programmability by carving out significant chunks of computation, while providing reuse across applications. In order to decompose applications into kernel-level accelerators, detailed workload characterization is needed to understand the speedup potential for different kernels.

Here we discuss some recent projects on hardware accelerators in each category of the taxonomy. The list of examples is not intended to be exhaustive.

2.2.1   ACCELERATORS THAT ARE PART OF THE PIPELINE.

One philosophy of integrating accelerators holds that if the accelerator is important enough, it should be put inside the pipeline as an execution unit. A classic example is the floating point unit (FPU). The advantage of integrating accelerators into the pipeline is that the accelerator design does not need to worry about the accelerator's interaction with the rest of the system: accelerators are just new functional units inside the pipelines, and they can leverage a core's load/store units and TLB for memory accesses. On the other hand, the performance of accelerators in this category can easily be limited by constraints of the cores, e.g., the bandwidth of the register files. These benefits and limitations are true for all the pipeline-attached accelerators. In this section, we have a representative accelerator design for each granularity in this category.

Instruction-Level

We start with instruction-level accelerators that are part of the pipeline.

FPU and SIMD.   In early Intel chips, there was no floating point unit. If the program required floating-point computing, programmers needed to emulate it in software, which was very slow. Eventually, with the increasing need for floating-point capability, floating-point processing was built into hardware. However, the first floating point unit is shipped as a co-processor on another chip (8087), which users could attach to the main CPU chip (8086) if they want. Not until the 80486 did Intel start integrating the floating-point unit with the CPU. Now the floating-point unit is a default component in a most processor core pipelines. Similar evolution also happened with SIMD units that speed up data-parallel computation.

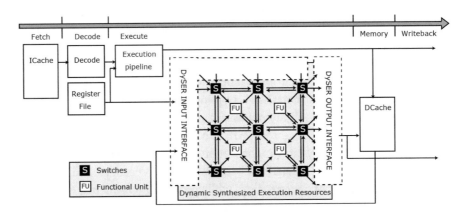

Figure 2.1: DySER pipeline [58].

DySER.   Dynamically Specialized Execution Resource (DySER) is a hardware-compiler co-design approach to dynamically wire a mesh of functional units for different phases of program execution. The key idea of DySER is to speed up commonly used computation paths without incurring repeated fetch, decode, and register access costs. DySER proposed an accelerator architecture integrated into the host processor's pipeline, shown in Figure 2.1, as a special execution unit. The DySER accelerator is composed of a heterogeneous array of functional units (FUs) connected with a mesh network. The granularity of the functional units in DySER is similar to the complexity of instructions in CPUs. Examples of FUs include integer ALU, multiply, and divide, as well as floating-point add/subtract, multiply, and divide [58].

Kernel-Level

Compared to instruction-level accelerators, kernel-level accelerators look at bigger chunks, e.g., multiple basic blocks, of computation.

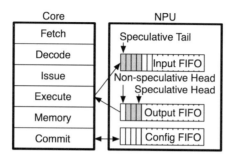

Figure 2.2: Neural processing unit [52].

NPU.   The neural processing unit (NPU) is a low-power accelerator of neural networks targeted at emerging approximate applications that can tolerate inexact computation in substantial portions of their executions [52]. The NPU is tightly coupled to the processor pipeline, shown in Figure 2.2, to accelerate small kernels, e.g., fft and sobel edge detector, in approximate applications. When a programmer writes their code for NPU, they explicitly annotate functions that are amenable to approximate execution. During compilation time, a NPU compatible compiler trains a neural network for the candidate region based on input-output training data, and generates codes that configures the NPU before the its invocation. The NPU configuration and invocation is done through ISA extensions that are added to the core.

10x10.   The 10x10 project envisions a customized but general-purpose architecture that exploits the benefits of customization for energy efficiency and performance, but maintains programmability and parallel scalability [40]. The goal of the project is to identify 10 most important kernels and then accelerate each kernel to be 10× more energy efficient and 10× faster [39]. The application coverage is achieved by optimizing ten distinct, but commonly used kernels. An example of the 10x10 architecture is shown in Figure 2.3. It has six kernel-level accelerators (micro-engines), including FFT, sort, and pattern matching (GenPM), and one RISC core. Each of the micro-engines is basically a specialized core with a customized functional units for the target kernels. New instructions are added to the ISA to invoke these specialized functional units.

Convolution Engine.   Convolution engines (CEs) are specialized for convolution-like dataflow kernels that are common in computational photography, image and video processing applications [98]. CE is developed as a specialized functional unit to Tensilica's extensible RISC cores [67]. Specific CE instructions are added to the ISA. The host RISC core decodes instructions in its instruction fetch unit and routes the appropriate control signals to CE if a CE instruc-

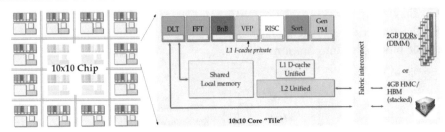

Figure 2.3:   An example of 10x10 architecture [40].

tion is encountered. The host core is also responsible for memory address generation, but the data is sent/returned directly from the internal register files in CE.

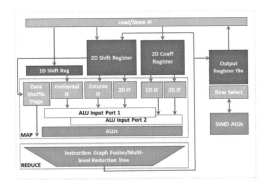

Figure 2.4:   Convolution engine system overview [98].

H.264 Accelerators.   A recent project from Stanford asked an interesting question: what are the sources of inefficiency in general-purpose processors [61]. The authors started from a 720p H.264 encoder running on a general-purpose processor, where the corresponding ASIC implementation is 500× more efficient than the general-purpose baseline. The paper then explores methods to eliminate general-purpose overheads by gradually transforming the CPU into a specialized system for H.264 encoding. Customized functional units, e.g., SIMD and fused operations, and customized storage, e.g., shift registers, are added to speedup important kernels inside the H.264 encoder. Similar to the Convolution Engine approach, this work also leverages the capabilities of Tensilica processors to add customized instructions into the host ISA. An example of the Tensilica processor is shown in Figure 2.5.

Application-Level
The most common application-level accelerators that are being integrated into the pipeline are cryptography accelerators [11, 18, 69]. Encryption/decryption algorithms are usually quite suitable for acceleration because of their high computational requirements and mature standard-

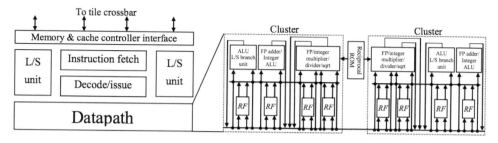

Figure 2.5: An example of the Tensilica processor pipeline [111].

ization. Implementations of the AES instruction set integrate the encryption/decryption accelerators into the processor pipeline and provide nearly an order of magnitude improvement in AES throughput. The Advanced Encryption Standard Instruction Set (AES-NI) is an extension of the x86 ISA for encryption and decryption using AES [18]. Oracle's SPARC and Cavium OCTEON II also include cryptography accelerators inside the core to accelerate security applications [11, 69].

2.2.2   ACCELERATORS THAT ARE ATTACHED TO CACHE

Both pipeline- and cache-coupled accelerators have unified address space with the host cores. However, an important distinction between them is that cache-coupled accelerators need to access the shared-coherent cache hierarchy without assistance from the host core's TLB or load-store unit. Typically, accelerators in this category also need to implement their own units for data movement and address translation.

Instruction-Level

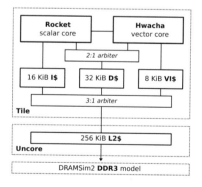

Figure 2.6: An example of Hwacha vector accelerators [95].

Hwacha. The Hwacha project is focusing on vector processing for data-parallel applications [76, 95, 121]. Figure 2.6 shows an example of a Hwacha accelerator. The Rocket scalar core in Figure 2.6 is a RISC-V core from UC Berkeley. A Hwacha accelerator is a vector co-processor that has its own instruction cache, but shares a data cache with the host core. As a good example of cache-coupled accelerators, a Hwacha accelerator has its own internal TLB for address translation and a load-store unit for cache access.

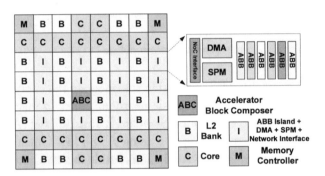

Figure 2.7: CHARM architecture [44].

CHARM and CAMEL. The Composable Heterogeneous Accelerator-Rich Microprocessor (CHARM) project focuses on composable fine-grained accelerator blocks that can be dynamically composed into accelerators for different workloads [44]. Figure 2.7 shows an overview of the CHARM architecture. CHARM is a heterogeneous architecture with cores, L2 caches, memory controllers, accelerator building block (ABB) islands (I), and an accelerator block composer (ABC) that dynamically connects different ABBs together for different functionalities. Inside each ABB is a dedicated scratchpad memory (SPM), a DMA engine, a NoC interface, and a set of fine-grained accelerators, like reciprocal, square-root, polynomial-16, and divide. CHARM supports a unified address space across the entire architecture, where a small, internal TLB is included in the DMA engine of each ABB. In the event of a TLB miss, the TLB request is forwarded to a shared TLB inside the accelerator block composer. If it also misses the shared TLB, the request is forwarded to the host core. CAMEL extends the CHARM architecture from ASIC to FPGA for better reconfigurability [43].

Kernel-Level

C-core, ECOcore, and QsCore. A series of efforts from the University of California, San Diego, focuses on energy-efficient, but not necessarily better performing, accelerators called Conservation Cores (C-cores) [102, 119, 120]. C-cores aim to efficiently execute hot regions of specific applications that represent significant fractions of the target system's workload. An example of C-core architecture is shown in Figure 2.8. The entire C-core tool chain starts with extracting the most frequently used code regions and synthesizing them into C-core hardware. In order to generate code for the C-core augmented processor, it extends a compiler framework with a combination of OpenIMPACT and GCC with knowledge of existing C-core on chip. The compiler uses a matching algorithm to find similarities between the input code and the C-core specifications. In the case of matching, the compiler generates a C-core-enabled binary that makes use of the C-core. The C-core accelerator shares the L1 cache of the host core, though the host CPU and C-cores do not simultaneously access the cache. The design assumes a coherent cache interface to enable communication between the C-core accelerator and the host core, but it does not mention how address translation is supported.

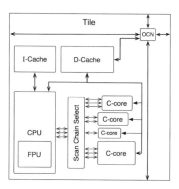

Figure 2.8: A C-core-enabled system [119].

SNNAP.   SNNAP is an FPGA prototype for neural network accelerators [91]. Instead of adding a neural network accelerator into a processor's pipeline, as in the case of NPU [52], SNNAP implements accelerators on an on-chip FPGAs, avoiding changes to the processor's ISA and microarchitecture. To program a SNNAP accelerator, application programmers can either use a high-level, compiler-assisted mechanism that automatically transforms regions of approximate

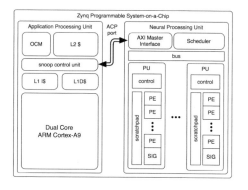

Figure 2.9: SNNAP [91].

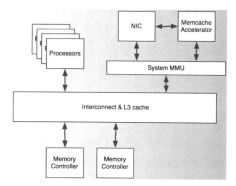

Figure 2.10: The overall Memcached architecture [80].

code to offload them to SNNAP accelerators, or a low-level, explicit interface that can batch multiple invocations together for pipelined processing. In the runtime, when a SNNAP-compatible program starts, it first configures the SNNAP accelerator with its topology and weights using the General Purpose I/Os (GPIOs) interface. Then, the program sends inputs through ARM Accelerator Coherency Port (ACP). The host processor then uses the ARMv7 SEV/WFE signaling instructions to invoke the SNNAP accelerator. The accelerator writes outputs back to the processor's cache via the ACP interface, and, when finished, signals the processor to wake up.

Application-Level

Memcached.   In-memory, key-value stores are an important component of modern data center services [80, 87]. Memcached is implemented using a hash table, with a unique key used to index the stored data. Figure 2.10 depicts a recent design of a memcached accelerator, called Thin Servers with Smart Pipes (TSSP). TSSP is designed for cost-effective, high-performance memcached deployment. It couples an embedded-class low-power core to a memcached accelerator that can process GET requests entirely in hardware. A system MMU translates virtual addresses to physical addresses that can be shared between the accelerators and the cores.

2.2.3   ACCELERATORS THAT ARE ATTACHED TO THE MEMORY BUS

Accelerators that are attached to the memory bus are usually coarser-grained accelerators since the invocation and offloading cost is more significant. Accelerators proposed in this category usually do not have much interaction with the host cores: they can access their own physical memory directly without worrying about address translation and coherence.

Kernel-Level

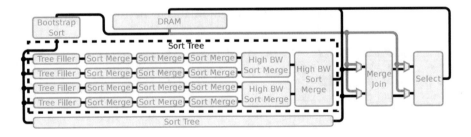

Figure 2.11: Hardware acceleration of database operations system overview [35].

Database Acceleration.   Database processing is an emerging area for acceleration. Its high memory bandwidth requirement makes it a perfect candidate for near-memory accelerators. Casper and Olukotun present an FPGA prototype using hardware to accelerate three important in-memory database primitives: selection, merge join, and sorting [35], all of which are connected directly to memory, as shown in Figure 2.11.

Tile	Area		Power		Critical Path
	mm^2	% Xeon [a]	mW	% Xeon	ns
Aggregator	0.029	0.07%	7.1	0.14%	1.95
ALU	0.091	0.21%	12.0	0.24%	0.29
BoolGen	0.003	0.01%	0.2	<0.01%	0.41
ColFilter	0.001	<0.01%	0.1	<0.01%	0.23
Joiner	0.016	0.04%	2.6	0.05%	0.51
Partitioner	0.942	2.20%	28.8	0.58%	***3.17
Sorter	0.188	0.44%	39.4	0.79%	2.48
Append	0.011	0.03%	5.4	0.11%	0.37
ColSelect	0.049	0.11%	8.0	0.16%	0.35
Concat	0.003	0.01%	1.2	0.02%	0.28
Stitch	0.011	0.03%	5.4	0.11%	0.37

Figure 2.12: Area/power/delay characteristics of Q100 tiles compared to a Xeon core [126].

Q100.   The Q100 is a Database Processing Unit (DPU) that can efficiently handle data-analytic applications [126]. It has a collection of heterogeneous fixed-function ASIC tiles, listed in Figure 2.12. Each of the tiles implements a database relational operator, such as a joiner or sorter, that resemble a common SQL operator. These tiles communicate with each other through an on-chip interconnect and load (store) data from (to) off-chip memory.

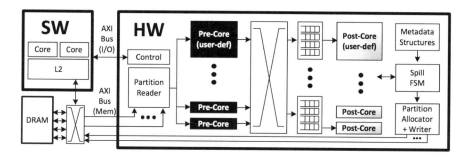

Figure 2.13: LINQits hardware overview [41].

LINQits.   The LINQits framework provides a pre-tuned accelerator template to accelerate a domain-specific query language, i.e., Language-Integrated Query (LINQ) [41]. LINQ supports a set of operators based on functional programming patterns and has been used to implement a variety of applications. Unlike other query languages, LINQ operators accept user-defined operators through a functional declaration as part of its processing. LINQits builds LINQ operators onto an SoC accelerator framework and provides a reconfigurable template to express user-defined functions. Figure 2.13 shows an overview of the LINQits framework. Data is streamed in from main memory through the partition reader, which is a specialized DMA engine responsible for reading bulk data.

Accelerator Store.   The Accelerator Store project proposed a shared memory framework for many-accelerator architectures. An accelerator characterization of a diverse pool of accelerators

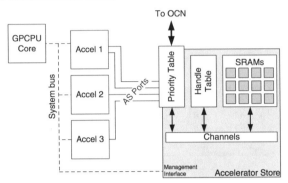

Figure 2.14: Accelerator store design [86].

reveals that each accelerator contains significant amounts of SRAM memory (40–90% of the total area) [86]. The characterization also shows that large private SRAM memories embedded within accelerators tend to have modest bandwidth and latency requirements. Figure 2.14 gives an example of accelerator store design. An accelerator store is a pool of SRAM memories connected through the memory bus. The pool can be shared by multiple accelerators. The handle table provides virtual memory support between them. An accelerator store can reduce the amount of on-chip memory for many-accelerator architectures with low overhead in terms of performance and energy.

Application-Level

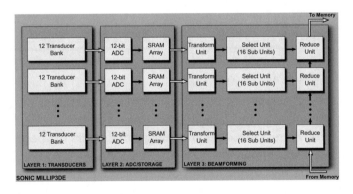

Figure 2.15: The Sonic Millip3De hardware overview [103].

Sonic3D. Figure 2.15 gives an overview of a standalone ASIC accelerator for 3D ultrasound beam formation—the most computationally intensive aspect of image formation [103]. The accelerator reads the input image from memory and writes the resulting updated image back to memory.

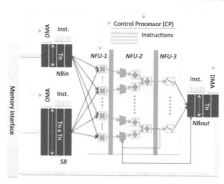

Figure 2.16: Diannao accelerator [37].

DianNao.   Machine learning is another emerging area where specialized accelerators could significantly improve performance and energy efficiency. One example of on-going projects in this area is the Diannao project [37, 38]. Diannao is an accelerator design for neural network algorithms, i.e., convolution neural networks (CNNs) and deep neural networks (DNNs), shown in Figure 2.16. DMA engines transfer data between accelerator buffers and memory.

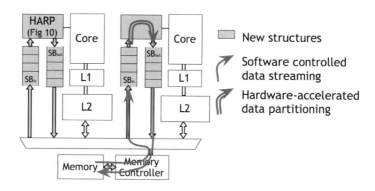

Figure 2.17: A 2-core system w/HARP integration [125].

HARP.   HARP is a hardware accelerator for range partitioning, which is central to modern database systems, especially for big-data analytics [125]. Figure 2.17 shows a block diagram of the major components in a system with HARP accelerators. The HARP accelerator is connected to the memory bus through two buffers; software moves data in and out of the buffers. A set of instructions is added to the host ISA for data movement orchestration between memory and HARP buffers. With data streaming in, the range partitioning is accelerated in the hardware accelerator.

2.2.4   ACCELERATORS THAT ARE ATTACHED TO THE I/O BUS

Here we move beyond the chip boundary: from on-chip accelerators to off-chip accelerators. A good example of programmable accelerators in this category is off-chip GPU with its own DRAM. Accelerators in this category is very loosely coupled with the host without much fine-grained communication. Usually significant speedup from the accelerator is expected to offset the communication cost.

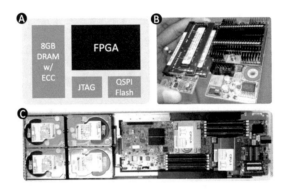

Figure 2.18: (a) Catapult FPGA block diagram. (b) Manufactured board. (c) The server that hosts the FPGA board. [97].

Catapult.   Catapult is a prototype FPGA-based accelerator for the Bing web search engine from Microsoft [97]. Catapult FPGAs, each with its own DRAM, are embedded into a half-rack of 48 machines. FPGAs are directly connected to each other in a 6x8 two-dimensional torus through general-purpose I/Os. Catapult is used to accelerate part of Microsoft Bing's ranking algorithm. When a server needs to rank a document, the software converts the document into a Catapult-compatible format and injects the document into its local FPGA. The FPGA pipeline computes the score for the document and sends the result back to the requesting server. Catapult was deployed with 1,632 servers running with mirrored Bing search traffic. Compared to a software-only implementation, Catapult achieves a 95% improvement in throughput with an equivalent latency distribution. On the other hand, it reduces the tail latency by 29% at the same throughput.

CHAPTER 3

Accelerator Design Flow 101.

" 'Tis a lesson you should heed:
Try, try, try again.
If at first you don't succeed,
Try, try, try again."
Popularized by William Hickson, 1836
(pre-dating the first work on High-Level Synthesis)

3.1 STANDARD RTL DESIGN FLOW

Despite the increasing popularity of hardware accelerators, the standard accelerator design flow is still very low-level and requires the use of complex and time-consuming electronic design automation (EDA) tools. Figure 3.1 illustrates the design flow for ASIC accelerators. It starts with a high-level description of an algorithm, then designers either manually implement the algorithm in Register-Transfer Level (RTL) using Verilog or VHDL or use high-level synthesis (HLS) tools, to compile the high-level programs to RTL. Hardware description languages such as Verilog and VHDL date back to the 1980's. They allow hardware designers to describe circuits using low-level building blocks, such as multipliers, registers, and multiplexers. Functional verification tests whether the resulting RTL design agrees with specifications. When all the blocks are implemented and verified, designers use commercial logic synthesis tools to map their designs to the gate level. Tools such as Synopsys Design Compiler take as input RTL in Verilog or VHDL, target tech-

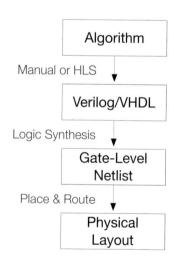

Figure 3.1: Synthesis flow.

nology libraries, and constraints, and they produce a gate-level netlist. This netlist is then transformed by place-and-route tools like Cadence SoC Encounter into the physical circuit layout. The whole process illustrated in Figure 3.1 is iterative, requiring a lot of tuning and refining at each stage.

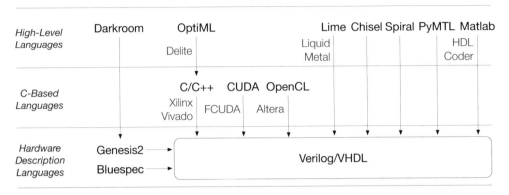

Figure 3.2: High-level synthesis landscape.

3.2 HIGH-LEVEL SYNTHESIS

Today's hardware designs are predominantly programmed using hardware description languages like Verilog and VHDL. Producing efficient designs using such inherently low-level languages requires a lot of expertise. It is time-consuming even for experienced hardware designers. Such a slow design process will not scale with tight SoC design cycles and the increasing use of hardware accelerators. Moreover, the low-level programming model also discourages application programmers, who have little hardware design knowledge, from converting applications into hardware. In addition, the history of software development in the past half century has demonstrated the value of higher abstraction levels to tackle growing complexity. For example, CPU programming benefits from well-established abstraction interfaces and advanced compilers, which free programmers from low-level details, increase their productivity, and reduce the likelihood of bugs.

To raise the level of abstraction in hardware designs, high-level synthesis tools that convert a high-level algorithm description into low-level RTL code are getting more attention in both academia and industry. The benefits of high-level synthesis include the following.

1. Better design. Higher abstraction levels free designers from worrying about low-level implementation details. With the help of a fast automation flow, designers can spend more time exploring alternative designs in terms of power, performance, and cost, potentially leading to better design choices.

2. Lower design cost. SoC companies have very tight design cycles. Greater automation in the design flow shortens design time and reduces expensive human involvement.

3. Greater accessibility for application designers. The future of specialization requires a lot of hardware-software co-design. Higher abstraction levels in hardware design could empower application programmers to more easily evaluate high-level algorithms in hardware.

Figure 3.2 summarizes the state-of-the-art high-level synthesis frameworks in both the commercial and research spaces. We divide high-level synthesis frameworks into three categories: hardware-description-language-based flow, e.g., Bluespec [94] and Genesis2 [106]; C-like-language-based flow, e.g., Xilinx Vivado [17], FCUDA [96], Altera SDK for OpenCL [1]; and high-level-language-based, e.g., Darkroom [63], OptiML/Delite [57, 113, 114], Lime/Liquid Metal [22, 23], Chisel [26], Spiral [46, 88], PyMTL [82], and Matlab HDL Coder [9]. The rest of this chapter aims to provide a pointer to some of the efforts in state-of-the-art tools. Interested readers can refer to the original papers/manuals to find more details.

3.2.1   BLUESPEC SYSTEMVERILOG

Bluespec SystemVerilog (BSV) is one of the early efforts to provide a higher-level hardware description language than Verilog/VHDL. Based on SystemVerilog syntax, Bluespec supports a higher level of abstraction than Verilog/VHDL in both behavioral and structural descriptions. Inspired by functional programming languages like Haskell, Bluespec supports more expressive types, overloading, encapsulation, and flexible parametrization to enable code reuse [94]. Programs written in Bluespec are compiled using the Bluespec compiler to generate corresponding RTL descriptions.

3.2.2   GENESIS2

Genesis2 is designed to create domain-specific hardware generators that encapsulate designer knowledge and only expose high-level application parameters. When running a generator, a user provides a set of parameters and constraints. From these, the generator produces the desired module. Instead of introducing a new hardware description language, Genesis2 extends SystemVerilog to exploit its verification support, and it uses Perl to provide flexible parametrization [106]. Darkroom [63] is an example of using Genesis2 to build image-processing pipelines in hardware.

3.2.3   XILINX VIVADO

Almost all CAD vendors have developed high-level synthesis tools, e.g., Cadence C-to-Slicing Compiler, Synopsys Synphony C Compiler, Mentor Graphics Catapult C, and Xilinx Vivado. Originally AutoESL, Vivado takes programs written in C/C++/SystemC as well as user-defined directives (similar to pragmas) and generates RTL. A subset of the C language is supported; features like recursive functions and dynamic memory allocation are prohibited.

As with other synthesis flows, the quality of Vivado-generated RTL designs is highly dependent on the input C code quality. Figure 3.3 demonstrates the quantitative effect that code quality can have on power and performance by comparing Pareto frontiers of optimized and unoptimized versions for the *Scan* benchmark. Both curves were generated by sweeping loop unrolling factors, memory bandwidth, and resource sharing, and by applying loop pipelining.

The unoptimized C code hits a performance wall at around 4000 cycles where neither increasing bandwidth nor loop parallelism yields better performance, despite consuming more

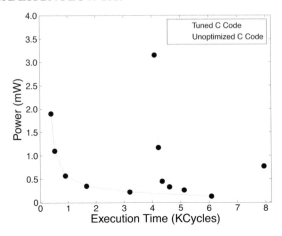

Figure 3.3: Unoptimized vs. tuned scan [109].

power. The reason is that when striding over a partitioned array that is both read from and written to in a given cycle, the HLS compiler conservatively adds loop-carried dependences, even though different array elements are being accessed. This in turn increases the iteration interval of loop pipelining, which limits performance. Partitioning the array differently simplifies access patterns, which allows HLS to resolve false dependences without violating conservative assumptions. This is typical of the tuning needed to produce efficient designs.

3.2.4   DELITE

There is increasing interest in domain-specific languages (DSL) for hardware compilation. DSLs can incorporate high-level, domain-specific features, and they can enforce restrictions that are not valid in general-purpose programming, making compiler analysis less conservative. Delite is a compiler framework that facilitates DSL development [114]. Delite simplifies DSL construction by providing common reusable components, like parallel patterns, optimizations, and code generations. Delite-generated DSLs, e.g., OptiML [113], are embedded in Scala, a general-purpose functional programming language [13], and can be compiled to multiple languages, like C++, CUDA, and OpenCL. Delite-generated C code, optimized with domain-specific knowledge conveyed in DSLs, can be used as input to C-based high-level synthesis tools, like Xilinx Vivado, to generate better RTL implementations [57].

3.2.5   LIME

The Liquid Metal project is a compiler and runtime system for heterogeneous architectures with a single and unified programming language [22]. This new programming language, called Lime, is intended to be executable across a broad range of architectures, from FPGAs to CPUs [23].

Based on Java, Lime has machine-independent semantics and a dynamic execution model, with extensions for parallelism, isolation and data flow. The compiler frontend performs optimization and compiles source code written in Lime to Java bytecode. The backend generates code for GPUs and FPGAs. The runtime dynamically picks the best implementation of a task based on available resources.

3.2.6   CHISEL

Chisel is a hardware construction language embedded in Scala [26]. Chisel includes a set of Scala libraries that define hardware datatypes and a set of routines to compile source code into either a cycle-accurate C++ simulator or a Verilog implementation. Examples of hardware built using Chisel include the RISC-V Rocket core [12], a key-value store accelerator [87], and the Hwacha vector accelerator [76].

3.2.7   SPIRAL

Spiral is a domain-specific hardware/library generator for signal processing [46, 88, 92]. In Spiral, a signal processing algorithm is first described as a set of formulas expressed in Spiral's signal processing language (SPL). Spiral recursively applies different transformations and optimizations to generate an optimal design based on program input and available hardware resources. Then the optimized representation is compiled into C or Verilog.

3.2.8   PYMTL

PyMTL aims to support computer architecture research by providing a vertically unified design environment for function-level, cycle-level, and RTL simulation [82]. Based-on Python 2.7, PyMTL allows designers to build architectural simulators at the function, cycle, or RTL level using a single language interface. For RTL simulation, PyMTL includes an RTL translator that translates the Python description of hardware into Verilog, which then is ported to the standard EDA design flow to produce power, area, and performance estimates.

CHAPTER 4

Accelerator Modeling

"Love truth, but pardon error."
Voltaire, 1738.

Computer architects have long been developing and leveraging high-level power [32, 79] and performance [24, 29] simulation frameworks for general-purpose cores and GPUs [27, 77]. In contrast, current accelerator-related research primarily relies on creating RTL implementations, a tedious and time-consuming process. It takes hours, if not days, to generate, simulate, and synthesize RTL to obtain the power and performance of a single accelerator design, even with the help of high-level synthesis (HLS) tools. Such a low-level RTL infrastructure cannot support architecture-level design space exploration that sweeps parameters across traditional general-purpose cores, accelerators, and shared resources such as cache hierarchies and on-chip networks. Hence, there is a clear need for a high-level design flow that abstracts RTL implementations of accelerators to enable broad design space exploration of next-generation customized architectures. This chapter focuses on Aladdin, a pre-RTL power-performance simulator designed to enable rapid design space search of accelerator-centric systems.

4.1 LIMITATIONS OF THE RTL-BASED DESIGN FLOW

The current accelerator design flow requires multiple CAD tools, which is inherently tedious and time-consuming. As described in Chapter 3, it starts with a high-level description of an algorithm, then designers either manually implement the algorithm in RTL or use HLS tools, such as Xilinx's Vivado HLS [17], to compile the high-level implementation (e.g., C/C++) to RTL. It takes significant effort to write RTL manually, the quality of which highly depends on designers' expertise. Although HLS tools offer opportunities to automatically generate the RTL implementation, extensively tuning C-code is still necessary to meet design requirements. After generating RTL, designers must use commercial CAD tools, such as Synopsys's Design Compiler and Mentor Graphics's ModelSim, to estimate power and cycle counts. If the generated design doesn't meet the power or performance specifications, part or even the whole process need to be repeated.

Despite the application-specific nature of accelerators, the accelerator design space is large given a range of architecture- and circuit-level alternatives. Figure 4.1 shows a large power-performance design space of accelerator design points for the GEMM workload from the SHOC benchmark suite [47]. The yellow square points were generated from a commercial HLS flow sweeping datapath parameters, including loop-iteration parallelism, pipelining, array partitioning, and clock frequency. However, HLS flows generally choose a fixed latency for all memory accesses, implicitly assuming local scratchpad memory fed by DMA controllers.

Such simple designs are not well suited for capturing data locality or interactions with complex memory hierarchies. The blue circle points in Figure 4.1 were generated by Aladdin integrated with a full cache hierarchy model and DRAMSim2 [101], sweeping not only datapath parameters but also memory parameters. By doing so, Aladdin exposes a rich design space that incorporates realistic memory penalties in terms of time and power, impractical with existing HLS tools alone.

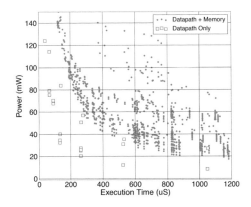

Figure 4.1: GEMM design space with and without memory hierarchy [109].

4.2   PRE-RTL MODELING—ALADDIN

Traditional low-level, RTL-based design flow cannot support architecture-level design space exploration that sweeps parameters across traditional general-purpose cores, accelerators, and shared resources such as cache hierarchies and on-chip networks. There is a clear need for a high-level design flow that abstracts RTL implementations of accelerators to enable broad design space exploration of next generation customized architectures.

This section will present Aladdin, a pre-RTL, power-performance-area simulation framework for hardware accelerators to address this need [109]. This framework takes high-level language descriptions of algorithms and accelerator design parameters as inputs, and outputs power, performance, and area, as well as cycle-level activities of accelerator implementations, including memory activity that can be fed to shared memory and interconnect models. Aladdin can be first used as an accelerator simulator that models an accelerator's behavior in an accelerator-rich SoC to facilitate the evaluation of the interaction between accelerators and shared resources like memory. Furthermore, Aladdin can also be used as an accelerator design assistant allowing SoC designers to quickly navigate the large design space of accelerators before they start producing RTL, greatly reducing design iterations.

The foundation of the Aladdin infrastructure is the use of dynamic data dependence graphs (DDDG) to represent accelerators. A DDDG is a directed, acyclic graph, where nodes represent computation and edges represent dynamic data dependences between nodes. The dataflow nature of hardware accelerators makes the DDDG a good candidate to model their behavior. Figure 4.2

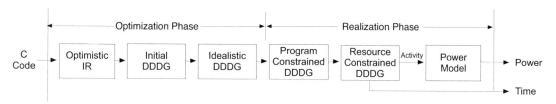

Figure 4.2: The Aladdin framework overview [109].

illustrates the overall structure of Aladdin, starting from a C description of an algorithm and passing through an *optimization* phase, where the DDDG is constructed and optimized to derive an idealized representation of the algorithm. The idealized DDDG then passes through a *realization* phase that restricts the DDDG by applying realistic program dependences and resource constraints. User-defined configurations allow wide design space exploration of accelerator implementations. The outcome of these two phases is a pre-RTL power-performance model for accelerators. Results show that Aladdin can model performance within 0.9%, power within 4.9%, and area within 6.6% compared to accelerator designs generated by traditional RTL flows. In addition, Aladdin provides these estimates over 100× faster.

Aladdin captures accelerator design trade-offs, enabling new architectural research directions in heterogeneous systems composed of accelerators, general-purpose cores, and the shared memory hierarchy seen in today's mobile SoCs and likely to be used in future customized architectures. Aladdin can model such systems because it has been integrated with the commonly-used general-purpose system simulator gem5 [29]. Such infrastructure allows users to explore customized and shared memory hierarchies for accelerators in a heterogeneous environment. In a case study with the GEMM benchmark, Aladdin uncovers significant high-level design trade-offs by evaluating a broad design space of the entire system. Such analysis results in more than 3× performance improvements compared to the conventional approach of designing accelerators in isolation.

Aladdin uses a DDDG to represent program behaviors so that it can take arbitrary C code descriptions of an algorithm—without any modifications—to expose algorithmic parallelism. This fundamental feature allows users to rapidly investigate different algorithms and accelerator implementations. Because of its optimiztic nature, dynamic analysis has been deployed previously in parallelism research exploring the limits of ILP [25, 54, 99, 122] and recent modeling frameworks for multicore processors [56, 68]. These studies sought to quickly measure the upper bound of performance achievable on an ideal parallel machine [72]. Aladdin has two main distinctions from these efforts. First, previous efforts model traditional Von Neumann machines where instructions are fetched, decoded, and executed on a fixed, but programmable architecture. In contrast, Aladdin models a vast palette of different accelerator implementation alternatives for the DDDG; the optimization phase incorporates typical hardware optimizations, such as removing memory operations via customized storage inside the datapath and reducing the bitwidth of functional units. The second distinction is that Aladdin provides a realistic power-performance model of accelerators across a range of design alternatives during its realization phase, unlike previous work that offered an upper-bound performance estimate [122].

In contrast to dynamic approaches, parallelizing compilers and HLS tools use program dependence graphs (PDG) [42, 55] that statically capture both control and data dependences [53, 59]. Static analysis is inherently conservative in its dependence analysis, because it is used for generating code and hardware that works in all circumstances, and because it is developed without run-time information. A classic example of this conservatism is the enforcement of false dependences that restrict algorithmic parallelism. For instance, programmers often use pointers to navigate arrays, and disambiguating these memory references is a challenge for HLS tools. Such situations frequently lead to designs that are more sequential than those that a human RTL programmer would develop. Therefore, although HLS tools offer the opportunity to automatically generate RTL, designers still need to extensively tune their C code to expose parallelism explicitly (Section 4.2.5). Thus, Aladdin is different from HLS tools; Aladdin is simply a realistic, accurate representation of accelerators, whereas HLS is burdened with generating actual, correct hardware.

The following sections describe details of Aladdin's optimization phase (Section 4.2.1) and realization phase (Section 4.2.2). Later discussion covers integration of Aladdin with memory systems (Section 4.2.3) and limitations of the approach (Section 4.2.4).

4.2.1   OPTIMIZATION PHASE

The optimization phase forms an idealized DDDG that only represents the fundamental dependences of the algorithm. An idealized DDDG for accelerators must satisfy three requirements: (a) only express necessary computation and memory accesses, (b) only capture true read-after-write dependences, and (c) remove unnecessary dependences in the context of customized accelerators. This section describes how Aladdin's optimization phase addresses these requirements.

Optimiztic IR

Aladdin builds the DDDG from a dynamic instruction trace, where the choice of the ISA significantly impacts the complexity and granularity of the nodes in the graph. In fact, a trace using a machine-specific ISA contains instructions that are not part of the program but produced due to artifacts of the ISA [108], e.g., register spills. To avoid such artifacts, Aladdin uses a high-level, machine-independent intermediate representation (IR) provided by the LLVM compiler infrastructure. LLVM IR is optimiztic because it allows an unlimited number of registers, eliminating additional instructions generated due to stack overheads and register spilling. The IR contains around 60 opcodes ranging from simple primitives, e.g., add and multiply, to complex operators, e.g., sine and square root, so that Aladdin can easily detect the functional units needed based on the program's IR trace and model them using pre-characterized hardware. A customized LLVM pass was developed to emit fully optimized IR instructions in a trace file. The trace includes dynamic instruction information such as opcodes, register IDs, parameter data types, parameter data values, and the dynamic addresses of memory operations.

Initial DDDG

Aladdin analyzes both register and memory dependences based on the IR trace. Only true read-after-write data dependences are respected in the initial DDDG construction. This DDDG is optimiztic enough for the purpose of ILP limit studies but is missing several characteristics of hardware accelerators; the next section discusses how Aladdin idealizes the DDDG further.

Idealized DDDG

Hardware accelerators have considerable flexibility to customize datapaths for application-specific features, which is not modeled in the initial DDDG. Such customization can change the attributes of the datapath, as in the case of bitwidth reduction where functional units can be tuned to the value range of the problem. Aladdin also removes operations that are not required for hardware implementations. For example, to reduce memory bandwidth, small, frequently accessed arrays, such as filters, can be stored directly in registers inside the datapath instead of in external memory. Cost models are used to automatically perform all of these transformations.

Aladdin optimizations can be categorized into node-level, loop-level, and memory-level transformations to produce an idealized DDDG representation.

Node-Level Optimization.   In addition to bitwidth analysis, Aladdin also models other node-level optimizations, such as strength reduction and tree-height reduction, by changing the nodes' attributes and performing standard graph transformations [65].

Loop-Level Optimization.   The initial DDDG captures true dependences between successive iterations of loop index variables, which means each index variable can only be incremented once per cycle. Such dependence constraints do not apply to hardware accelerators or parallel processors since it is possible that they can initiate multiple iterations of a loop simultaneously [115]. Aladdin removes all dependences between loop index variables, including basic and derived induction variables, to expose loop parallelism.

Memory Optimization.   The goal is to remove unnecessary load/store operations. In addition to the memory-to-register conversion example described above, Aladdin also performs store-load forwarding inside the DDDG, which eliminates load operations by buffering data in internal registers within hardware accelerators. This is different from store-load forwarding in general-purpose CPUs, where the load operation must still be executed [105].

Extensibility.   Hardware design is open-ended, and Aladdin can be extended to incorporate other accelerator-specific optimizations, analogous to adding new microarchitectural structures to CPU simulators. A CAM is an example of a custom circuit structure that is often used to accelerate hash tables in network routers and datatype-specific accelerators [131]. Unlike software, CAMs can automatically compare a key against all of the entries in one cycle. On the other hand, large CAMs are power hungry, resulting in an energy trade-off when hash tables reach a certain size. Aladdin incorporates CAMs into its customization strategy by automatically replacing software-managed hash tables with CAM. Aladdin can detect a linear search for a key by looking for chained sequential memory look-ups and comparison.

4.2.2   REALIZATION PHASE

The realization phase uses program and resource parameters, defined by users, to constrain the idealized DDDG generated in the optimization phase.

Program-Constrained DDDG
The idealized DDDG optimiztically assumes that hardware designers can eliminate all control and false data dependences at design time. Aladdin's realization phase models actual control and memory dependences to create the program-constrained DDDG.

Control Dependence.   The idealized DDDG does not include control dependences, assuming that branch outcomes can be known in advance and operations can start before branches are resolved, which is unrealistic even for hardware accelerators. The costs and benefits of control flow speculation for accelerators have not been extensively studied yet, and one solution for minimizing control dependences relies on predicated execution to simultaneously execute both taken and not taken paths until branch resolution [73]. While this approach minimizes serialization, the cost of speculation is very high—it requires hardware resources that grow exponentially with the number of outstanding branches. Aladdin models control dependence by bringing code from the not-taken path into the program-constrained DDDG to account for additional power and resources. Aladdin is flexible enough to model the costs of different mechanisms for handling control flow. For energy efficiency, Aladdin models one outstanding branch at a time, serializing control dependences for multiple simultaneous branches.

Memory Dependence.   The idealized DDDG optimiztically removes all false memory dependences between dynamic instructions, keeping true read-after-write dependences. This is realistic for memory accesses with addresses that can be resolved at design time. However, some algorithms have input-dependent memory accesses, e.g., histogram, where different inputs result in different dynamic dependences. Without run-time memory disambiguation support, designers have to make conservative assumptions about memory dependences to ensure correctness. To model realistic memory dependences, the realization phase includes memory ambiguation that constrains the input-dependent memory accesses by adding dependences between all dynamic instances of a load-store pair, as long as a true memory dependence is observed for any pair. This is similar to the dynamic dependence profiling approach adopted by parallelization efforts [56, 70].

Resource-Constrained DDDG

Table 4.1: Realization phase user-defined parameters, $i::j::k$ denotes a set of values from i to k by a stepping factor j

Parameters	Example Range
Loop Rolling Factor	[1::2::Trip count]
Loop Pipelining	On or Off
Clock Period (ns)	[1::2::16]
FU latency	Single-Cycle, Pipelined
Memory Ports	[1::2::64]

Finally, Aladdin accounts for user-specified hardware resource constraints, a subset of which are shown in Table 4.1. Users specify the type and size of hardware resources in an input configuration file. Aladdin then binds the program-constrained DDDG onto the hardware resources, leading to the resource-constrained DDDG. Aladdin can easily sweep resource parameters to explore the design space of an algorithm, which is fast because only resource constraints need to be applied for each design point. These resource parameters are set with respect to the following three factors: loop rolling, loop pipelining, and memory ports.

Loop Rolling.   The optimization phase removes dependences between loop index variables, assuming completely unrolled loops that execute all iterations in parallel. In reality, for loops with large trip counts, this leads to large resource requirements. Aladdin's loop rolling factor re-rolls loops by adding dependences between loop index variables.

Loop Pipelining.   The DDDG representation fully pipelines loop iterations by default, though sometimes pipelined implementation leads to high resource requirements as well as high power consumption. Aladdin offers users the option to turn off loop pipelining by adding dependences between the entry and exit nodes of successive loop iterations.

Memory ports.   The number of memory ports constrains the data transfer rate between the accelerator datapath and the closest memory hierarchy, generally either a scratchpad memory or L1 cache. Aladdin uses this parameter to abstractly model the number of memory requests the datapath can issue concurrently. Section 4.2.3 describes how the memory ports interface with memory simulators.

An Example

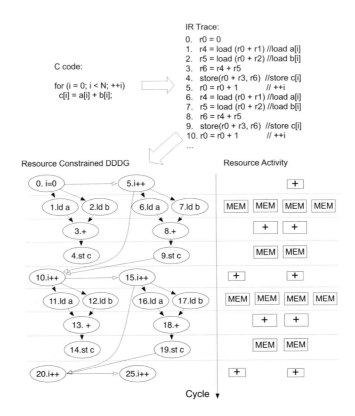

Figure 4.3: C, IR, resource constrained DDDG, and activity [109].

Figure 4.3 illustrates different phases of Aladdin transformations using a microbenchmark as an example. After the IR trace of the C code has been produced, the optimization and realization phases generate the resource-constrained DDDG that models accelerator behavior. In this example, the user wants an accelerator with a factor-of-2 loop-iteration parallelism and without loop pipelining. The solid arrows in the DDDG are true data dependences, and the dashed arrows represent resource constraints, such as loop rolling and turning off loop pipelining. The horizontal dashed lines represent clock cycle boundaries. The corresponding resource activities are shown to the right of the DDDG example. Such a constrained DDDG reflects the dataflow nature of the accelerator. Aladdin can accurately capture the dynamic behavior of accelerators without having to generate RTL by carefully modeling the opportunities and constraints of the customized datapath in the DDDG.

Power and Area Models

This section describes the construction and application of Aladdin's power and area models to capture the resource requirements of accelerators.

Power Model.   To accurately model the power of accelerators, Aladdin needs: (a) precise activities and (b) accurate power characterization of different DDDG components. Aladdin uniquely characterizes switching, internal, and leakage power from Design Compiler for each type of DDDG node (multipliers, adders, shifters, and so on) and registers. The characterization accounts for different timing requirements, bitwidths, and switching activity. Switching and internal power are due to capacitive charging/discharging of output load and internal transistors of the logic gates, respectively. While switching and internal power are both dynamic, internal power is weakly dependent on activity because internal nodes can switch without causing the gate output to switch.

Aladdin constructs its power model by synthesizing microbenchmarks that exercise the functional units. The microbenchmarks cover all of the compute instructions in IR so that there is a one-to-one mapping between nodes in the DDDG and functional units in the power model. These microbenchmarks are synthesized using Synopsys's Design Compiler in conjunction with a commercial 40 nm standard cell library to characterize the switching, internal, and leakage power of each functional unit. This characterization is fully automated in order to easily migrate to new technologies.

Aladdin's power modeling library also accounts for cell selection variances during netlist synthesis. Different pipeline stages within a datapath contain varying amounts of logic and, in order to meet timing requirements, different standard cells and logic implementations of functional units are often selected at synthesis time. Aladdin approximates the impact of cell selection by training the model for a variety of timing constraints and using a first-order model to choose the correct design. This also accounts for the logic flattening that Synopsys's Design Compiler performs across small collections of functional units.

Area Model.   To accurately model area, Aladdin also includes an area library similar to the previously described power library for each DDDG component. This model was obtained using the same set of microbenchmarks to characterize the area for each functional unit as well as for registers.

Cycle-Level Activity.   Figure 4.4 shows the cycle-level resource activity for one implementation of the FFT benchmark. Aladdin accurately captures the distinct phases of FFT. The number of functional units required is estimated using the maximum number of parallel functional units per cycle for each program phase; this approximation provides the power and area models with the total resources allocated to the accelerators. The cycle-level activity is an input to the power model to represent the dynamic activity of the accelerators.

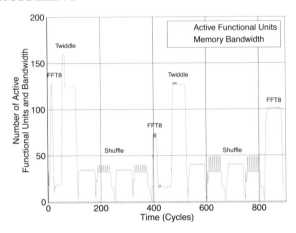

Figure 4.4: Cycle-by-cycle FU and memory activity of FFT [109].

4.2.3   INTEGRATION WITH MEMORY SYSTEM

Aladdin can easily integrate with architectural cache and memory simulators to model behavior with a particular memory hierarchy. Within the context of memory hierarchy for accelerators, Aladdin supports three types of integrated memory models.

Ideal Memory guarantees that all memory requests can be serviced in one cycle, which is only realistic for a system with a small memory size. Aladdin models the ideal memory system by assuming load and store nodes in the DDDG take one cycle.

Scratchpad Memory is commonly used in accelerator-centric systems where accelerator designers explicitly manage memory accesses so that each request has a fixed latency. However, this approach requires a detailed understanding of workload memory characteristics. This potentially increases design time but leads to more efficient implementation. Aladdin can take a parametrized memory latency as an input to model the latency of load and store operations matching the characteristics of scratchpad memory.

Cache Hierarchy applies a hardware-managed cache system to capture the locality of the accelerated workload. Such a cache hierarchy relies on the hardware to exploit the locality of the workload, potentially easing the design of systems with a large number of accelerators. On the other hand, a cache introduces variable memory latency. Aladdin enables cache simulators to be integrated to evaluate how variable-latency memory accesses affect accelerator behaviors.

In order to integrate with a cache hierarchy, the accelerator must include certain mechanisms to react to possible cache misses. Aladdin models several approaches to handling this variable latency, which resemble pipeline control mechanisms in general-purpose processors. The simplest policy is local or global pipeline stalls on miss events. A more complex mechanism for non-blocking behavior could be that a new loop iteration is started when a miss occurs, and only the loop ID is stored for re-execution when the miss resolves.

Memory Power Model.   The memory power model is based on a commercial register file and SRAM memory compiler that accompanies the standard cell library. Aladdin also includes a memory power model from CACTI [123]. For consistency, the results discussed in this section use the memory power model from the standard cell library.

4.2.4   LIMITATIONS

Algorithm Choices.   Aladdin does not automatically sweep different algorithms. Rather, it provides a framework for quickly exploring various hardware designs of a particular algorithm. This means designers can use Aladdin to quickly and quantitatively compare the design spaces of multiple algorithms for the same application to find the most suitable algorithm choice.

Input Dependence.   Like other dynamic analysis frameworks [66, 83], Aladdin only models operations that appear in the dynamic trace, which means it does not instantiate hardware for code not executed with a specific input. For Aladdin to completely model the hardware cost of a program, users must provide inputs that exercise all paths of the code.

Input C Code.   Aladdin can create a DDDG for any C code, but when modeling accelerators, Aladdin does not deal with C constructs that require resources outside the accelerator, such as system calls and dynamic memory allocation. In fact, understanding how to handle such events is a research direction that Aladdin facilitates.

4.2.5   ALADDIN VALIDATION

This section describes the traditional RTL design flow and workloads used to validate Aladdin. Validation results show Aladdin has modest error rates, within 0.9% for performance, 4.9% for power, and 6.5% for area. Aladdin generates the design space more than 100× faster than the traditional RTL-based flow.

Figure 4.5 outlines the methodology used to validate Aladdin. Aladdin's power and area estimates are compared against synthesized Verilog generated by Design Compiler using commercial 40nm standard cells. Aladdin's performance model is validated against ModelSim Verilog simulations. The SAIF activity file generated from ModelSim is fed to Design Compiler to capture switching activity at the gate level. The Verilog designs are either hand-coded or generated using Xilinx's Vivado HLS tool. The RTL design flow is an iterative process and requires extensive tuning of both RTL and C code.

A collection of benchmarks, implemented either by hand or using HLS, are used to validate Aladdin. HLS enabled the validation of the Pareto optimal designs for the SHOC benchmarks, overcoming the impracticality of hand coding each design point. Aladdin was validated against

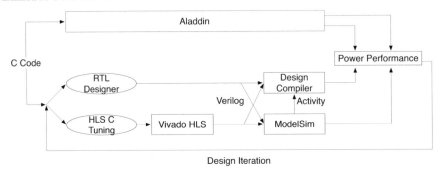

Figure 4.5: Validation flow [109].

handwritten RTL for benchmarks ill-suited for HLS. Examples are taken from recently published accelerator research: NPU [52], Memcached [80], and HARP [125].

Figure 4.6 shows that Aladdin accurately models performance, power, and area across all of the chosen benchmarks, with average deviation of 0.9%, 4.9%, and 6.5%, respectively, from the RTL implementations. For each SHOC workload, we validated six points on the Pareto frontier. The SHOC validation results show that Aladdin accurately models entire design spaces, while for single accelerator designs, Aladdin is not subject to the shortcomings of HLS and can accurately model different customization strategies.

Pareto Analysis.   The Pareto optimal designs of the SHOC benchmarks reveal interesting program characteristics in the context of hardware accelerators. Bars in Figure 4.10 correspond to six designs along each benchmark's Pareto frontier, which were also used for validation. In each graph, the leftmost bar is the most parallel, highest performing design while the rightmost bar is the most serial and lowest performing design. For each design, energy is calculated using power and performance estimates from Aladdin. Aladdin's detailed power model enables energy breakdowns for adders, multipliers, and registers. The six bars of each benchmark are normalized to the leftmost bar to facilitate comparisons.

Each of the three benchmarks in Figure 4.10 exhibits different energy trends across the Pareto frontier. Triad, shown in Figure 4.7, demonstrates good energy proportionality, meaning more parallel hardware leads to better performance with a proportional power cost. In contrast, Sort has a strong sequential component such that energy increases for more parallel designs without improving performance. Finally, while the multiplier energy for Stencil shows similar energy proportionality to Triad, the adders and registers required for loop control are amortized with more parallelism. Non-intuitively, this leads to better energy efficiency for these faster designs.

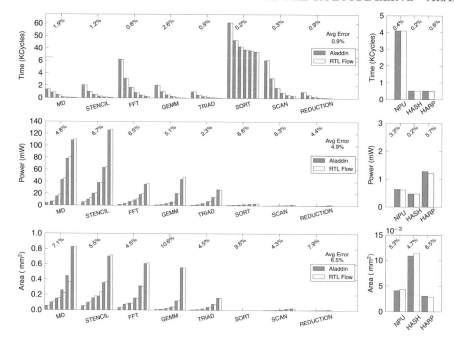

Figure 4.6: Performance (top), Power (middle), and Area (bottom) Validation [109].

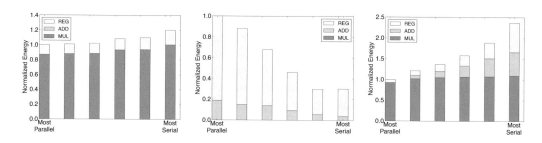

Figure 4.7: Triad.   Figure 4.8: Sort.   Figure 4.9: Stencil.

Figure 4.10: Energy characterization of SHOC [109].

4.2.6   ALGORITHM-TO-SOLUTION TIME

Aladdin enables rapid design space exploration of accelerator designs. Table 4.2 quantifies the differences in algorithm-to-solution time to explore a design space of the FFT benchmark with 36 points. Compared to traditional RTL flows, Aladdin skips the time-consuming RTL generation, synthesis, and simulation process. On average, it takes 87 min to generate a single design using the RTL flow but only 1 min for Aladdin, including both Aladdin's optimization phase (50 s) and

Table 4.2: Algorithm-to-solution time per design

	Hand-Coded RTL	HLS	Aladdin
Programming Effort	High	Medium	
RTL Generation	Designer Dependent	37 min	N/A
RTL Simulation Time	5 min		
RTL Synthesis Time	45 min		
Time to Solution per Design	87 min		1 min
Time to Solution (36 designs)	52 h		7 min

its realization phase (12 s). However, because Aladdin only needs to perform the optimization phase once for each algorithm, this optimization time can be amortized across large design spaces. Consequently, it only takes 7 min to enumerate the full design space with Aladdin compared to 52 h with the RTL flow. The HLS RTL generation time per design is comparable to that reported by other researchers [81].

4.2.7 CASE STUDY: GEMM DESIGN SPACE

This section presents a case study that demonstrates how Aladdin enables architecture research and why it is invaluable for future heterogeneous SoC designs.

The case study touches on several important aspects of accelerator design.

1. Execution Time Decomposition: Understanding design trade-offs of an accelerator's execution time with respect to *compute time* and *memory time*.

2. Accelerator Design Space: Characterizing accelerator design space, including memory hierarchy, to understand how different parameters affect the design space.

3. Heterogeneous SoC: Demonstrating the impact of a single accelerator on resource contention in an SoC-like system, resulting in different optimal designs that would be unknown without system-level analysis.

Execution Time Decomposition

So far, Aladdin has been evaluated as a standalone accelerator simulator with an ideal memory hierarchy (one cycle memory access latency). However, it is not always possible to retrieve data in one cycle in real designs with large problem sizes. The efficiency of accelerators highly depends on the memory system. To quantify the impact of a memory system on accelerators, Aladdin is integrated with a standard cache simulator and the DRAMSim2 memory simulator [101].

The accelerator's execution time is divided into *compute time* and *memory time*. Compute time is defined as the execution time of an accelerator when memory latency is only one cycle.

Memory time is defined as cycles lost to a non-ideal memory with realistic memory latency constraints.

In order to decompose the accelerator's execution time, both an ideal memory and a realistic memory hierarchy including L1, L2, and DRAM simulations are compared. The compute time is the execution time with ideal memory latency; the difference in execution times between the two simulations is the memory time required to move data into the accelerator [33].

Table 4.3: Single accelerator design space, where $i::j::k$ denotes a set of values from i to k by a stepping factor of j [109]

Type	Parameters	Values
Algorithm	Blocking Factor	[16, **32**]
L1	L1 Bandwidth (Bytes/Cycle)	[4::2::128]
	L1 Size (KB)	[4::2::32]
	MSHR Entries	[4::2::64]
L2	L2 Bandwidth (Bytes/Cycle)	[4::2::128]
	L2 Size (KB)	[64::2::256]
	L2 Assoc	16

Table 4.3 lists all of the parameters in the design space. This section focuses on the bandwidth and size of L1. Figure 4.11 shows the execution time and power breakdown of the GEMM benchmark when sweeping L1 size and bandwidth. The left of Figure 4.11 demonstrates that memory time takes a significant portion of the execution time, especially as L1 bandwidth increases memory latency becomes more dominant. With the same L1 bandwidth, execution time decreases as the L1 size increases from 8 KB to 16 KB; this phenomenon occurs because 8 KB is not large enough to hold the blocked data size (a 32×32 matrix).

The plot on the right shows the power breakdown of the accelerator datapath, L1, and L2. The accelerator datapath power increases with L1 bandwidth, because higher bandwidth enables more-parallel implementations. As L1 size increases, its power also increases as accesses become more expensive. At the same time, L2 power decreases because more accesses are coalesced by the L1, lowering the L2 cache's activity. In fact, cache power accounts for more than half of the total power, even for more parallel designs where datapath power is significant. Therefore, design efforts focusing on the accelerator datapath alone do not alleviate memory power, which dominates the overall power cost.

Accelerator Design Space

Section 4.2.7 explored a subset of the design space for accelerators and memory systems. This section demonstrates the use of Aladdin to explore the comprehensive design space with parameters in Table 4.3. Figure 4.12 plots the power and execution time of the GEMM accelerator designs resulting from the exhaustive sweep. The design space contains several overlapping clusters of similar designs. The arrows in Figure 4.12 identify correlations in power/performance trends with

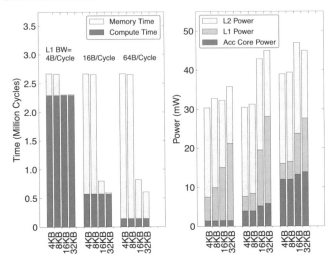

Figure 4.11: GEMM time and power decomposition [109].

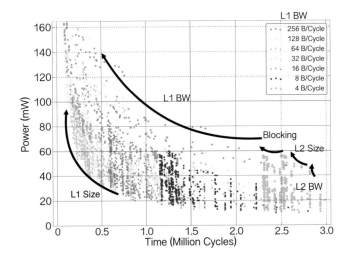

Figure 4.12: GEMM design space [109].

respect to each parameter. For example, GEMM experiences substantial performance benefits from a larger L1 cache, but with a significant power penalty. In contrast, increasing L2 size only modestly increases both power and performance.

Resource-Sharing Effects in Heterogeneous SoC

In a heterogeneous system, shared resources, such as a last-level cache, can be accessed by both general-purpose cores and accelerators. As an example, let us consider the case of a heterogeneous system consisting of a shared 256 KB L2 cache, one general-purpose core, and a GEMM accelerator with a private 16 KB L1 cache. From an accelerator designer's perspective, an important algorithmic parameter is GEMM's blocking factor. A larger blocking factor exposes more algorithmic parallelism, but achieving good locality requires a larger cache.

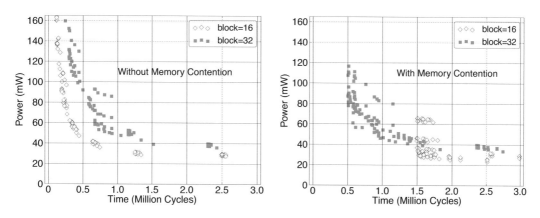

Figure 4.13: Design space of GEMM without and with contention in L2 cache [109].

Figure 4.13 (left) shows the accelerator design space without memory contention from the general-purpose core. The algorithmic blocking factor has been swept, and the figure shows that a blocking factor of 16 is always better than 32 with respect to both power and performance. This occurs because a 16 KB L1 cache is large enough to capture the locality of blocking factor 16 but not 32. Therefore, it is preferable to build the accelerator with blocking factor 16 when there is no contention for shared resources.

To model resource contention between the general-purpose core and the accelerator, Pin[84] is used to produce an x86 memory trace and then use the trace to issue requests that pollute the memory hierarchy while simultaneously running the accelerator. The design space for the accelerator under contention is shown in Figure 4.13 (right). Performance degrades for both blocking factors of 16 and 32 due to pollution in the L2 cache; however, blocking factor 32 suffers much less than blocking factor 16. When there is contention, capacity misses increase for the shared L2 cache, which incurs large main memory latency penalties. With a larger blocking factor, the accelerator requires fewer references to the matrices in total and, thus, fewer data requests from the L2 cache. Consequently, the effects of resource contention suggest building an accelerator with a larger blocking factor, where the accelerator performance can achieve around 0.5 million cycles. This example shows the importance of modeling the entire system, not just standalone components. Without considering the contention, designers may pick a design with blocking factor 16, the highest performance of which is 1.5 million cycles in the contention scenario. Such a design choice leads to a $3\times$ performance degradation, compared to the system using a blocking factor of 32. Aladdin can easily evaluate these types of system-wide accelerator design trade-offs, a task that is not tractable with other current accelerator design tools.

CHAPTER 5

Workload Characterization for Accelerators

"If you know the enemy and know yourself, you need not fear the result of a hundred battles. If you know yourself but not the enemy, for every victory gained you will also suffer a defeat. If you know neither the enemy nor yourself, you will succumb in every battle." Sun Tzu, *The Art of War*.

Accelerators are intrinsically tailored to applications, and workload characterization plays a large role in developing these architectures. Tuning an architecture toward a workload requirement demands a comprehensive understanding of the intrinsic characteristics of the workload. This chapter is about workload characterization for accelerator design. We will introduce WIICA, an Instruction Set Architecture (ISA)-independent workload characterization tool.

5.1   ISA-INDEPENDENT WORKLOAD CHARACTERIZATION—WIICA

Workload characterization for general-purpose architectures is commonly done by profiling benchmarks on current generation microprocessors using hardware performance counters. Typical program characteristics are machine instruction mix, IPC, cache miss rates, and branch misprediction rates. This approach is limited because machine-dependent features such as cache size and pipeline depth strongly affect the characterization of the workload. To overcome this problem, *microarchitecture-independent* workload characterization can be employed by profiling instruction traces to collect information such as working set sizes, register traffic, memory locality, and branch predictability [64]. Although this approach removes the effects of microarchitecture-dependent features, some of these analyses depend on the particular ISA used to collect the trace. Each ISA has different characteristics and constraints that impact the representation of the workload. As architectural specialization grows in importance, *ISA-independent* workload characterization will become essential for understanding intrinsic workload behavior, which will in turn allow designers to consider a wide range of alternative architectures.

To fully expose not only microarchitecture-independent but also ISA-independent workload characteristics for specialized architectures, an approach called Workload ISA-Independent

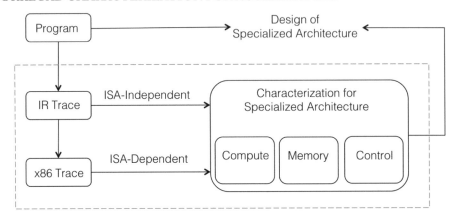

Figure 5.1: WIICA overview.

Characterization for Applications (WIICA) has been proposed for analyzing benchmarks using ISA-independent characteristics that capture inherent program behavior (see Figure 5.1). To enable this analysis, WIICA leverages the ISA-independent nature of an existing compiler intermediate representation (IR). WIICA uses a JIT compiler to trace workloads using this ISA-independent program representation and to compare program characteristics within the broad categories of program compute, memory activity, and control flow. In particular, WIICA studies program characteristics that are highly relevant to the design of specialized architectures. In this section, we discuss how to use WIICA to do workload characterization for accelerators, and we illustrate the difference between ISA-independent and ISA-dependent characterization.

5.1.1   WHY ISA-INDEPENDENT?

Accelerators are unburdened by the requirements of legacy ISAs, and much of the efficiency gained by using accelerators can be attributed to hardware specialization of the datapath, memory, and program control. ISA-independent analysis is attractive for such architectures because it avoids artificial constraints imposed by details of a specific ISA. These constraints affect the behavior of programs because compilers for conventional ISAs must generate binaries that meet the specification of the instruction set semantics. This subsection discusses the effects of three major kinds of ISA constraints: the overhead of stack operations caused by register spilling, ISA-specific complex operators, and calling conventions.

Stack Overhead.   The set of registers defined by an instruction set architecture is necessarily smaller than or equal to the set of physical registers in a machine. Most high-level language code uses variables liberally, without regard for the number of registers in a particular target system. In order to fit the large number of variables into the ISA-defined register set, compilers must perform register allocation to map program variables to registers. When there are more live variables needed than available architectural registers, the compiler spills some of them onto the stack, a special area of main memory. Load and store operations are inserted to move spilled values into and out of machine registers for computation. These stack memory operations can be expensive from a run-time performance point of view. When characterizing workloads for specialized architectures that do not have a fixed or known ISA, such stack accesses add irrelevant load/store operations to the instruction trace that distort memory utilization.

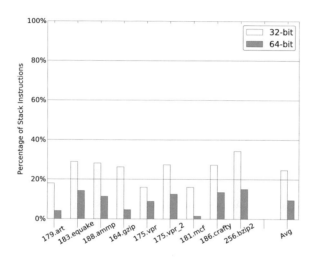

Figure 5.2: The percentage of stack instructions of total dynamic instructions for 32-bit and 64-bit x86 binaries.

To demonstrate the effect of stack operations, we compare 32-bit and 64-bit x86 binaries generated by LLVM's Clang compiler for a set of SPEC CPU benchmarks. One of the major differences between the 32- and 64-bit x86 ISAs is that the 64-bit x86 ISA has eight more general-purpose registers. Figure 5.2 plots the percentage of dynamic instructions that access the stack for 32-bit and 64-bit implementations of the benchmark set. Note that for each benchmark, the 32-bit version executes a much higher percentage of stack instructions than the 64-bit version. Additional general-purpose registers allow more variables to stay in registers, so less spilling to memory is required.

The stack overhead also applies to RISC ISAs. Lee et al. characterized stack access frequency using the Alpha ISA to propose a mechanism for separating stack from heap accesses [75]. For the same SPEC CPU2000 workloads, they find a percentage of stack operations (24%) that is similar that in Figure 5.2 for 32-bit x86.

Complex Operations.   Two classes of instructions can be categorized as complex operations: vector instructions and compute or branch instructions with memory operands. Both kinds of operations can be split into multiple simpler primitives. CISC ISAs like x86 contain complex operations including vector instructions like SSE and instructions that support memory operands. However, complex operations can exist even in RISC ISAs. For example, POWER and ARM include complex operations such as predicate instructions, string instructions, and vector extensions.

When we perform workload characterization for specialized architectures, it is easier and cleaner to start from simple primitives and explore aggregation possibilities rather than to start from a more complex version of code resulting from another category of optimization.

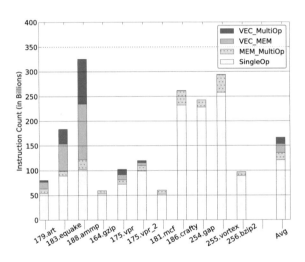

Figure 5.3: Instruction breakdown of complex (top three bars) and single (bottom bar) operation instructions.

Figure 5.3 quantifies the amount of complex operations in x86. In this categorization, an instruction is classified as a complex operation if it is either a vector instruction (SSE) or a compute or branch instruction with a memory operand. The top three categories in Figure 5.3 are complex operations: vector operations, vector operations with memory accesses, and compute or branch instructions with memory operands. The remaining category includes all single operation instructions. We see that on average 27% of the total instructions executed are complex operations.

Calling Convention.   The ISA calling convention describes how subroutines receive parameters from callers and how they return results. Any machine-dependent ISA needs to have its own specifications for passing arguments between subroutines. For example, due to its limited number of registers, x86 pushes all arguments onto the stack before a subroutine is called, resulting in additional stack operations. Other ISAs also require various housekeeping operations for subroutines, and these are also artifacts of the ISA choice, not intrinsic to the behavior of the workload. These additional instructions, mostly stack operations, due to calling conventions could potentially be misleading for specialized architectures that do not have the specific calling convention.

5.1.2   METHODOLOGY AND BACKGROUND

To evaluate the importance of performing workload characterization by using machine-independent code representation, we perform both ISA-independent and ISA-dependent analysis.

ISA-Independent Study.   An ISA-independent representation of code is critical for the development of flexible compiler infrastructures. Fortunately, modern compilers use ISA-independent intermediate representations to bridge high level source languages (e.g., C) to specific ISAs (e.g., Intel x86). Since our requirements for code representation are similar to those of compilers, WIICA leverages the intermediate representation used in compilers to perform its analysis.

WIICA uses the intermediate representation (IR) in both LLVM [8] and ILDJIT [34].[1] Specifically, workload execution is represented by a trace of semantically equivalent IR instructions, instead of ISA-dependent binaries. This dynamic IR instruction trace is generated by executing the IR code with a special-purpose interpreter that emits dynamic IR instructions as it executes them.

Compiler's IR.   ILDJIT is a modular compilation framework that includes both static and dynamic compilers [34]. ILDJIT performs a large set of classical, machine-independent optimizations at the IR level including copy propagation, dead-code elimination, loop-invariant code motion, and the like. When the IR code is fully optimized, it is translated to LLVM's bitcode language and LLVM's back ends are used to optimize the code using machine-dependent optimizations and to generate semantically equivalent machine code.

We customized ILDJIT to implement an ad-hoc interpreter of its intermediate representation that emits IR instructions as they are executed. The IR instructions interpreted are the ones used for translation to the bitcode language. By attaching our interpreter right before the translation to bitcode, we ensure that the IR is fully optimized; however, machine-dependent information is still not used for these optimizations, allowing our analysis to study workload-specific characteristics.

The ILDJIT IR is a linear machine- and ISA-independent representation that includes common operations of high-level programming languages like memory allocation (e.g., new, free, newarray) and exception handling (e.g., throw, catch). It is a RISC-like language in which memory accesses are performed through loads and stores. Each instruction has a clear and simple meaning; only scalar variables, memory locations, and the program counter are affected by its

---

[1]We use the results obtained with ILDJIT IR for the discussion but the methodology can be extended to LLVM IR as well. The current WIICA distribution uses LLVM IR because the ILDJIT framework is no longer being actively developed.

execution. The language allows an unbounded number of typed variables (virtual registers), making analysis independent of the number of physical registers. Moreover, parameters of method invocations are always passed by using variables, as in the input source language we use (C), making analysis independent of specific calling conventions. Finally, the data types described in the source language are preserved in the IR language, making this representation closer to the input language compared to other compiler intermediate representations. Consequently, the three ISA-dependent concerns we are studying - register spilling, complex instructions, and calling conventions - will not appear in the IR being produced.

IR instructions that perform operations among variables require homogeneity among their types: an add operation between variables x and y requires the same type for both x and y (e.g., 32-bit integer). This characteristic leads to instructions that convert values between types. Notice that these conversions are required by the workload as the semantics of operations in the source language specify them. However, some of these conversions are unnecessary if a CISC-like ISA is used instead of the RISC-like IR. Finally, opcodes (e.g., `add`, `mul`) are orthogonal with data types (e.g., integer, floating point). This opcode polymorphism constrains the number of different instructions in the language to 80, allowing an easy parsing of the executed trace.

ISA-Dependent.   To demonstrate the difference between ISA-dependent and -independent analysis, we use the x86 instruction set for the ISA-dependent analysis. The x86 ISA is commonly used in architecture studies, and many program analysis tools are available for workload characterization. For analysis of new x86-based microarchitectures, architects must understand the ISA-specific effects of the architecture since they can have a significant impact on pipeline and memory system design. When considering new heterogeneous architectures with both x86 and specialized cores, it would be natural to use existing workload characterization approaches. However, when performing workload characterization of specialized architectures, our results show that x86 provides a particularly poor starting point. In this study, we compare x86 instruction traces with IR traces. To generate the trace of x86 instructions executed by a workload, we use Pin, a dynamic binary instrumentation tool developed by Intel [85].

Sampling.   Because of storage and processing time constraints, applying some of the analysis presented in this chapter to a full execution trace is impractical. Therefore, we sample execution with SimPoint[110]. We configure SimPoint to generate 10 phases, each of which contains 10 million instructions. Only instructions that belong to the identified phases are emitted and then analyzed.

In order to make a fair comparison between x86 and IR traces, we sample the execution of the IR trace by configuring SimPoint to use IR instructions. Then we instrument the code to identify the x86 instructions semantically equivalent to the IR code for the identified phases. In this way, we ensure that the same code region is considered for both the IR and x86 analysis.

Benchmark Suite.   We use C benchmarks from SPEC CPU2000 benchmark suite. These benchmarks are translated to CIL bytecode by the compiler GCC4CLI [5] (a branch of GCC), and then they are compiled to IR by ILDJIT. Finally, ILDJIT generates the machine code by relying on LLVM's x86 back end as previously described. ILDJIT currently only supports the 32-bit LLVM back end and all of the results in this section are for 32-bit operations.

5.1.3   COMPUTE

Accelerators often exploit custom functional units that combine multiple operations with pre-dictable control flow in order to execute code more efficiently. An example of this approach is conservation cores [119], which identifies the hot functions in a program's execution and designs hardware accelerators for those functions. In order to uncover the opportunity to find sequences of operations that are amenable to similar specialization, executed instruction sequences need to be analyzed to detect specific patterns. For such analysis, the way the operations are represented in the instruction trace will have a significant impact on whether certain patterns can be found or not and, subsequently, whether the workload is worth the effort of custom hardware design. In this section, we analyze the instruction breakdown and the most common opcodes found in both x86 and IR. We observe that x86 incurs more overhead for the basic computation performed by the application.

Instruction Breakdown.   We start the analysis by categorizing the executed instructions from the IR and x86 code. We split instructions into the following categories: Stack, Memory, Move (data movement and conversion between registers), Unconditional Branch, Conditional Branch, and Compute. Figure 5.4 shows this breakdown. For each benchmark, the leftmost bar represents the x86 binary, and the middle bar represents IR. Furthermore, during our implementation, we found that there is also instruction overhead associated with IR characteristics that are not intrinsic to the workloads. One source of such inefficiency is the number of unconditional branch instruc-tions. The ILDJIT compiler does not remove these instructions because the compiler back end performs unconditional branch removal in a very efficient manner. Another source of overhead is data movement and conversion between registers. Such instructions appear in both IR and x86 and are used to support different data types and to simplify optimizations. The rightmost bar in Figure 5.4 is what we call Simplified-IR—the IR trace without those two classes of instruction. In the following discussion, "IR trace" will refer to this simplified IR.

     As with the results from LLVM's 32-bit Clang compiler in Figure 5.2, we see that, de-pending on the application, the number of stack-referencing instructions can be significant. This is represented by the top section of the leftmost bar for every benchmark. For example, almost half of the x86 instructions for 255.vortex use the stack, while the effect is less obvious for bench-marks like 179.art. More importantly, the large number of stack accesses results from constraints imposed by the x86 ISA (a small register set) and not from the intrinsic behavior of the pro-gram. This is evident from the IR bars for the stack-heavy benchmarks—moving from x86 to the infinite-register IR significantly decreases the number of accesses to the stack. While stack effects can increase the number of executed x86 instructions, CISC x86 instructions can combine multiple primitive operations together. This results in a more compact execution. For example,

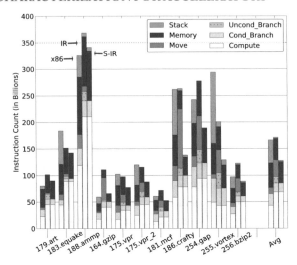

Figure 5.4: The instruction breakdown for x86, IR and Simplified-IR (S-IR).

for benchmarks like 164.gzip and 179.art there are more instructions in the IR trace than in the x86 one. The presence of x86-specific effects that both increase and decrease executed instructions makes it even harder to assess ISA-dependent overhead and expose the workload's intrinsic behaviors, further strengthening the case for analysis on the IR level.

Opcode Diversity.   Our next experiment examines the diversity of the opcodes in the x86 and IR traces. Opcode diversity is relevant since it is related to the complexity of customized functional units in specialized hardware. Fewer and simpler opcodes will simplify the design of such hardware because the functional units will be more modular and reusable. This allows sharing such functional units across various workloads.

In order to compare x86 and IR analysis, we profile the total number of opcodes and the number of times each single opcode occurs in the program execution. We do not differentiate opcodes based on addressing modes, which reduces the number of required x86 opcodes. Figure 5.7 plots the number of unique opcodes and the percentage of dynamic instructions those opcodes cover for the benchmark 179.art. The dotted line on the plot shows the cumulative distribution of opcodes needed to cover the dynamic execution of the program.

To meaningfully compare x86 and IR, we use a horizontal line to highlight the number of unique opcodes required to cover 90% of the dynamic instructions in Figures 5.5 and 5.6. This metric is meaningful for accelerator studies since it allows comparison of the number of functional unit types needed for different workloads. The horizontal line intersects with the cumulative distribution function to show the required number of opcodes. The x86 results demonstrate that 90% of the execution can be covered by 12 unique opcodes, while the same analysis with IR requires

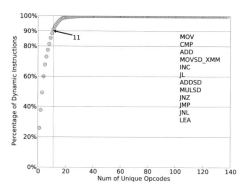

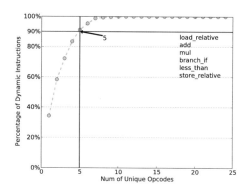

Figure 5.5: x86   .

Figure 5.6: IR

Figure 5.7: Cumulative distribution of the number of unique opcodes of 179.art. The intersecting lines show the number of unique opcodes that cover 90% of dynamic instructions.

only 6 opcodes (the X axis of Figures 5.5 and 5.6 starts from 0). The right portions of the plots show the top opcodes used for both instruction sets. For x86, two MOV instructions, MOV and MOVSD_XMM, and four different conditional jump instructions are required. Compared with x86, the top opcodes from IR analysis are much clearer—the 6 opcodes are all simple primitives, resulting in a much simpler representation of the actions of the program.

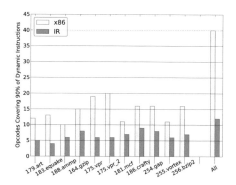

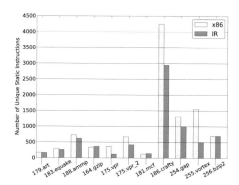

Figure 5.8: Number of unique opcodes to cover 90% of dynamic instructions. "All" represents the global superset.

Figure 5.9: Number of unique static instructions to cover 90% of dynamic instructions.

We extend this comparison to all available benchmarks in the suite and show the result in Figure 5.8. Not surprisingly, for each benchmark the x86 trace needs more unique opcodes than the IR trace. Furthermore, the rightmost bar in Figure 5.8 shows the number of unique opcodes required to cover all benchmarks we analyze, computed as a superset of individual benchmark needs. In order to cover all the benchmarks in x86, 40 unique instruction opcodes are required, but the IR-based analysis uncovers only 12 fundamental primitives. Thus, extracting workload pieces that are amenable to hardware specialization appears significantly easier on the IR level of abstraction.

Static Instructions.   The diversity of opcodes represents the different types of fundamental computing blocks that custom hardware might require. Another important metric is the number of static instructions (or the size of the executable code) required to cover the dynamic execution. In a custom design, different sequences of static instructions will lead to more or less complex data flow. As with the metric we use for opcode analysis, we compare the number of unique static instructions required to cover 90% of the dynamic instructions. Figure 5.9 shows that benchmarks that have significant stack overhead (shown in Figure 5.4), like 186.crafty and 255.vortex, require more unique static instructions, most of which are potentially stack operations. In that case, the x86 characterization hides the truly important instructions, instead highlighting the stack overhead operations. This may skew the identification of hot computation.

5.1.4   MEMORY

Memory behavior is crucial for workload performance. In the case of hardware specialization, the memory system must be tuned to the workload characteristics in order to realize significant gains in efficiency. In this section, we compare two memory characterization metrics, memory footprint size and memory entropy. We once again discover that ISA-dependent analysis can be significantly misleading and obscure the workloads' intrinsic behavior.

Memory Footprint.   The first metric we consider is the size of the data memory that a program uses, including both stack and heap memory. We look into two types of memory footprint. The first one is the full memory footprint—the total size of data memory the program has accessed. It quantifies the overall memory usage. The second metric identifies the "important" memory footprint, which we define as the number of unique memory addresses that cover 90% of dynamic data memory accesses. This metric shows the most frequently used addresses that need to be kept close to the computation.

Figure 5.10 shows the total memory footprint analysis. The Y-axis in this figure is the number of unique memory addresses generated. The x86 and IR memory footprints are nearly the same, because the total working set is intrinsic to the workloads and therefore independent of the program representation.

However, the important memory footprints of x86 and IR, shown in Figure 5.11, are markedly different. In most cases, fewer unique memory addresses are needed for x86 than for IR, despite the fact that their total numbers of unique memory addresses are similar. The reason for this once again lies in frequent accesses to the stack. While the memory space of stack addresses is usually small, these addresses are accessed very frequently. When identifying important memory addresses, the few stack addresses that are frequently accessed will stand out and dominate the memory behavior. Thus, the important memory addresses found will be an artifact of the ISA instead of the program behavior.

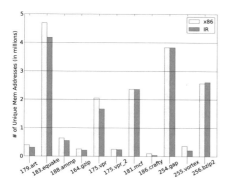

Figure 5.10: Number of unique memory addresses to cover 100% of dynamic memory accesses.

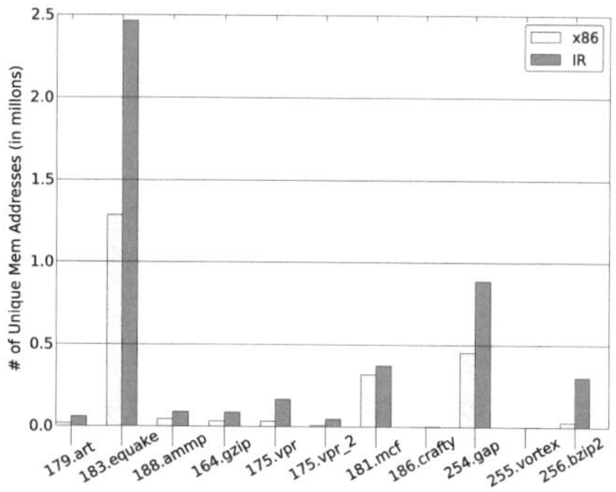

Figure 5.11: Number of unique memory addresses to cover 90% of dynamic memory accesses.

Memory Address Entropy.   We introduce *memory address entropy* as a metric that quantifies how easy it is to keep memory data close to the computation. Intuitively, memory address entropy reflects the uncertainty, the lack of predictability, of data addresses accessed by a workload. Thus, it is a metric opposite to memory locality, which is often exploited by custom hardware. Locality measures the regularity of a memory address stream, while entropy measures its lack. We show that ISA-level analysis exposes a lower amount of entropy, leading to false assumptions of memory access regularity.

Entropy.   In information theory, entropy [107] is used to measure the randomness of a variable, which is calculated as in Equation 5.1:

$$Entropy = -\sum_{i=1}^{N} p(x_i) * \log_2 p(x_i), \tag{5.1}$$

where $p(x_i)$ is the probability of x_i, and N is the total number of samples of the random variable x. The result, $Entropy$, is a measure of predictability of the next outcome of x. For example, assume the pattern of variable x is very regular—always 1. In this case, $p(1) = 1$ and $\log_2 p(1) = \log_2 1 = 0$, so $Entropy = 0$, which means that it is very easy to predict x. On the other extreme, if there are N possible outcomes of x occurring equally often, $p(x_i) = \frac{1}{N}$. According to Equation 5.1,

$$Entropy = -\sum_{i=1}^{N} p(x_i) * \log_2 p(x_i)$$
$$= -N * \frac{1}{N} * \log_2(\frac{1}{N})$$
$$= \log_2 N$$

which is very high for large N.

Yen et al. describe the idea of using entropy to represent the randomness of instruction addresses[127]. According to Equation 5.1, in the case of memory entropy, variable x represents the memory addresses that appear in the program execution. The probability $p(x_i)$ is the frequency of a specific memory address x_i. After profiling the unique memory addresses accessed in the workloads and the number of times each address is referenced, we can compute the memory address entropy of the workloads. When the memory entropy is high, the memory access stream is more random and less amenable to architecture techniques that require locality. Conversely, if the entropy is low, memory accesses are very regular and easier to predict.

Global Memory Address Entropy.   Global memory entropy describes the randomness of the entire data address stream using all address bits (32 in our case). Figure 5.12 shows the calculated global memory address entropy for both x86 and IR. For each benchmark, the leftmost bar is the global entropy of x86 memory addresses; the rightmost bar is that of IR memory addresses. We can see that the entropy of the x86 address stream is generally much lower than that for IR, meaning better temporal locality. In order to find the reason for the difference, we compute the x86 memory address entropy without the stack addresses, shown in the middle bar. As we can see, after removing the stack addresses, the x86 address entropy is now comparable with the IR memory address entropy. So the higher temporal locality shown in the x86 trace is mostly due to the presence of stack operations. There are two major reasons. First, most of the stack references are from spilled variables. These spilled variables do not exist in IR's memory trace because they are preserved as register operations. Second, it is entirely possible that variables in different phases of the program are mapped to the same stack address. In this case, accesses to these different variables all seem to fetch the same stack address from the x86 memory trace, leading to higher locality. Such locality is completely an artifact of the ISA, not representing the intrinsic locality of programs.

Local Memory Address Entropy.   Local memory entropy computes the address entropy using a subset of the address bits. Local entropy can help detect spatial locality in the workloads. For example, we can skip the lower-order bits of the addresses and compute entropy only with the high-order address bits, as seen in Figure 5.13. If the local address entropy with 28 bits, for example, shrinks significantly compared to global entropy, then memory accesses are less random,

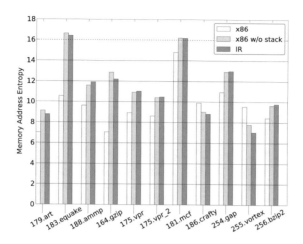

Figure 5.12: Memory address entropy of x86, x86 without stack, and IR traces. Lower values indicate more regularity in the access stream.

and significant spatial locality is present. Ignoring the lower order bits reveals spatial locality by grouping those addresses that are close together.

Figure 5.16 shows two examples of the local address entropy when we sweep the number of low-order bits ignored from 0–10. The two benchmarks, 179.art and 255.vortex, are representative of the patterns we have seen among the rest of the benchmark suite. For both cases, the local entropy of x86 drops faster than for IR. This is very obvious for 255.vortex. This is due to the fact that stack addresses are usually in close proximity, which means they have usually have good locality. Ignoring the lower-order bits results in steeper drops in entropy for x86. This also shows ISA-dependent analysis will bias the workload characteristics toward better locality due to the impact of stack operations.

5.1.5   CONTROL

Control flow complexity is an important metric for workload characterization. From experience with general purpose processor design, we know that speculative execution is necessary to exploit parallelism. In a heterogeneous architecture, there may be a variety of cores or computing engines with different degrees of support for speculation. In order to choose the appropriate ones to run the workloads, the control complexity of the workloads needs to be fully understood and not dependent on a specific architecture. In this section, we compare the control complexity analysis of x86 and IR and show that the two analyses are consistent with each other, showing that ISA choice has a minimal effect on a workload's control flow.

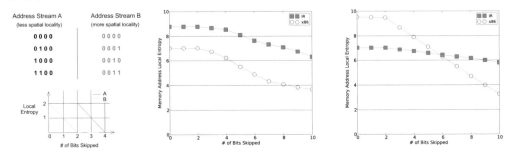

Figure 5.13: Example of local entropy.   Figure 5.14: 179.art   Figure 5.15: 255.vortex

Figure 5.16: Local memory entropy as a function of low-order bits omitted in calculation. A faster dropping curve indicates more spatial locality in the address stream.

Branch Instruction Count.   Our first-order control flow analysis counts the number of unique conditional branch instructions that cover 90% of the dynamic branches. This is similar to the unique opcode analysis but focuses on branch instructions. This is important for hardware specialization because it measures the number of control flow decisions that must be handled in a design.

As Figure 5.17 shows, the number of unique branch instructions required to cover 90% dynamic branches is consistent between x86 and IR. The two sets of bars track each other very well. This implies that ISA choice does not have a significant impact on the number of branch instructions generated, which mostly depends on the way programs are written.

Branch History Entropy.   Another important metric is control flow predictability, which is intrinsic to the workload. Generally speaking, if the branch taken patterns are more regular and less random, branches are easier to predict. In this sense, the regularity of the branch behavior will indicate the predictability of the control flow. Based on this intuition, Yokota proposed the idea of *branch history entropy* using Shannon's information entropy idea to represent a program's predictability [128].

We use a string of bits to encode taken or not taken branch outcomes. In this sense, the program, as the producer of the sequence, can be viewed as an information source and we can compute the entropy of the information source to represent the regularity of branch behavior. In our implementation, we use a sequence of n consecutive branch results as the random variable and compute the entropy of the benchmarks. The results are shown in Figure 5.18. We can see that the branch entropy from x86 and IR also track each other very well. This shows that both ISA-dependent analysis and ISA-independent analysis fully expose the program's control behavior. This matches our intuition that ISA does not affect control flow significantly.

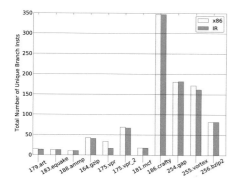
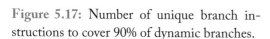

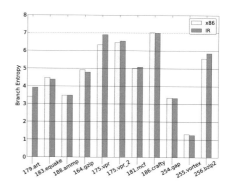

Figure 5.17: Number of unique branch instructions to cover 90% of dynamic branches.

Figure 5.18: Branch entropy per workload. Lower values imply better branch predictability.

5.1.6   PUTTING IT ALL TOGETHER

We compare the eleven SPEC benchmarks with five ISA-independent metrics from our analysis: the number of opcodes, the value of branch entropy, the value of memory entropy, the unique number of static instructions (I-MEM), and the unique number of data addresses (D-MEM). In terms of specialized architecture design, smaller values for each of these metrics indicate more regularity in the benchmarks and better opportunity to exploit specialization. For each metric, we choose the maximum value across all the benchmarks and for each benchmark we plot the relative value with respect to this maximum value. Figure 5.19 shows kiviat plots for all benchmarks, in which each axis represents one of the ISA-independent characteristics. The plot in the lower right corner of the figure provides a legend for the individual axes. The kiviat plots are ordered by the area of the resulting polygon. With an equal weighting of the five characteristics, area provides a rough approximation for overall benchmark regularity (smaller area is more regular). We observe very different behavior across the benchmark suite. For example, 255.vortex demonstrates regularity across all the metrics, while 186.crafty has relatively low regularity in most of the dimensions. These insights will be helpful for specialized architecture designers to identify the opportunities for acceleration.

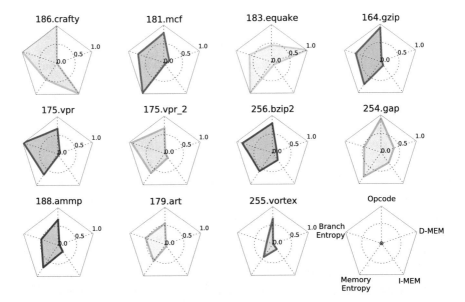

Figure 5.19: Comparison of five ISA-independent metrics across SPEC benchmarks, ordered by the area of the polygon. The lower right kiviat plot provides the legend, and smaller values indicate more regularity in the metric.

CHAPTER 6

Accelerator Benchmarks

"A foolish consistency is the hobgoblin of little minds, adored by little statesmen and philosophers and divines." Ralph Waldo Emerson, *Self-Reliance*, 1841.

The quantitative nature of computer architecture design has made benchmarking an important part of the design process. Over the years, a wide variety of benchmarks have been proposed to help researchers evaluate new architecture designs. As we enter the era of hardware accelerators, designers need to understand how best to use available benchmarks to explore hardware accelerator design. This chapter provides a taxonomy of benchmarks and uses it to classify today's benchmark suites.

Table 6.1: Benchmark taxonomy

Architecture-Specific		Domain-Specific	
Architectures	**Benchmarks**	**Domains**	**Benchmarks**
CPU	SPEC CPU2000/2006 [14, 15]	Data Mining	MineBench [93]
CMP	PARSEC [28] SPLASH-2 [124]	Multimedia	MediaBench [74] ALPBench [78]
GPU	Rodinia [36] Parboil [112] SHOC [47]	Approximate Computing	AxBench [3]
Embedded	Mibench [60]	Medical Image	UCLA Medical Image Pipeline [45]
Accelerator	MachSuite [100] CHStone [62]	Sensing	SDVBS [118] CortexBench [116]

Table 6.1 provides a taxonomy covering existing benchmark suites. In general, there are two themes in benchmark design. Some benchmark suites are designed to evaluate a specific kind of architecture. For example, the SPEC CPU suites [14, 15] are for CPUs and Rodinia [36] and SHOC [47] are for GPUs. We call these *architecture-specific* benchmark suites. Suites in this category usually include quite diverse workloads, covering different application behaviors, so that they can be used to evaluate how robust a new architectural feature is across a diverse set of workloads.

The second approach is to design a benchmark suite for a specific application domain, like MineBench [93] for data mining and SDVBS [118] for vision. We call these *domain-specific* benchmark suites. Unlike application-specific benchmark suites, domain-specific benchmark suites are not designed for broad coverage of application characteristics. Instead, they aim to provide a set of benchmarks that are representative enough for a specific domain. Domain-specific benchmarks developed by the architecture community are always a reflection of an emerging application domain that demands more computational power than existing hardware can provide.

Benchmarks for Accelerators

Today's accelerator projects show a great disparity in workload selection. A detailed survey of recent high-level synthesis and accelerator-related papers by Reagen et al. revealed that of the 88 distinct benchmarks used across 25 papers, 64 of them were only used *once* [100]. In general, programmable accelerators tend to use general-purpose benchmarks to demonstrate their coverage. For example, the evaluation of DySER uses benchmarks from SPEC, PARSEC, and Parboil [58]. Application-specific accelerators focus on specific applications or classes of applications, like data partitioning for HARP [125] and 3D ultrasound image formation for Sonic3D [103].

To improve standardization in accelerator evaluation, one of the recent efforts in accelerator benchmarking is MachSuite, an accelerator-centric benchmark suite tailored to the needs of accelerator research. MachSuite is a set of 19 benchmarks spanning 12 kernels, written to cover a diverse set of application domains and to incorporate multiple algorithmic choices [100]. MachSuite was also designed with high-level synthesis in mind. All of the benchmarks in MachSuite are HLS synthesizable, providing architecture researchers an easy way to quickly generate a diverse set of hardware accelerators. Table 6.2 lists the benchmarks in MachSuite with a short description of each and the application space it covers.

Table 6.2: The MachSuite benchmarks [100]

Kernel/Algorithm	Description	Berkeley Dwarf [21]
AES/AES	AES encryption	Combinational logic
BACKPROP/BACKPROP	Neural network training	Unstructured grids
BFS/BULK	Breadth-first search	Graph traversal
BFS/QUEUE	Breadth-first search	Graph traversal
FFT/STRIDED	Fast Fourier transform	Spectral methods
FFT/TRANSPOSE	Fast Fourier transform	Spectral methods
GEMM/NCUBED	Matrix multiplication	Dense linear algebra
GEMM/BLOCKED	Matrix multiplication	Dense linear algebra
KMP/KMP	String matching	Finite state machines
MD/KNN	Molecular dynamics	N-body methods
MD/GRID	Molecular dynamics	N-body methods
NW/NW	DNA alignment	Dynamic programming
SORT/MERGE	Sorting	Map reduce
SORT/RADIX	Sorting	Map reduce
SPMV/CRS	Sparse matrix/vector multiplication	Sparse linear algebra
SPMV/ELLPACK	Sparse matrix/vector multiplication	Sparse linear algebra
STENCIL/STENCIL2D	Stencil computation	Structured grids
STENCIL/STENCIL3D	Stencil computation	Structured grids
VITERBI/VITERBI	Hidden Markov model estimation	Graphical models

CHAPTER 7

Future Directions

"If your plan is for one year, plant grain. If your plan is for ten years, plant trees. If your plan is for one hundred years, educate children. "
Guan Zhong, Spring and Autumn Period.

Specialized architectures have been a growing topic in both academic research and commercial development for the past decade. As traditional technology scaling slows, specialization becomes a viable solution for computer architects to continue performance growth and energy efficiency improvements without relying on technological advances.

In this book, we discussed the driving force for recent interest in specialization and the proliferation of research proposals from computer architecture and design automation communities on accelerator designs. In particular, we introduced two research tools, Aladdin and WIICA, for accelerator modeling and characterization. We also discuss new benchmark suites available to evaluate accelerator architectures. We hope this book serves as an introduction to the field, as an overview of the vast literature on accelerator designs, and as a resource for researchers working in related areas.

For example, researchers can use the benchmarks from Chapter 6 as inputs to workload characterization tools in Chapter 5 to recognize potential opportunities for acceleration. Based on workload characteristics, researchers could explore potential accelerator architectures based on the accelerator taxonomy from Chapter 2. Once candidates for acceleration are identified, the modeling tools in Chapter 4 could be used to explore accelerator and system configurations and optimization. If a particular design looks promising, the high-level synthesis tools discussed in Chapter 3 could then be used to realize the design.

A number of challenges remain to be addressed in future work. Here we highlight three major challenges in the field of accelerator architectures as our community moves forward.

1. Flexibility. Specialized accelerators, especially fixed-function accelerators, are only designed for specific applications or domains of applications. *How should we choose a combination of different accelerable kernels to achieve a good balance between application coverage and energy efficiency?* Composing accelerators and general-purpose cores can address this issue if we can understand the common kernels across a group of applications and identify efficient communication channels to chain these kernels so that they can work together to achieve greater functionality.

2. Design Cost. The increasing volume and diversity of accelerators in every generation of processors requires rolling out new designs quickly with relatively low design cost. RTL-based implementations through standard flows are inherently tedious and time-consuming. High-level synthesis tools have shown promise to speedup the design process. However, existing tools still face challenges to generate high-quality designs within reasonable time. *How can we rapidly create and validate new accelerators with good quality of design?* To achieve this, we must develop new program representations, compilation heuristics, and algorithms that can aid high-level synthesis tools to quickly generate high-quality designs. Domain-specific high-level synthesis approaches also are a promising direction.

3. Programmability. Programming modern high-performance SoC is similar to the state of GPGPU programming before the introduction of language extensions like CUDA and OpenCL. Prior to the introduction of those languages, leveraging GPUs for general-purpose computing was only possible for experienced high-performance computing or game programmers. Similarly, the sophisticated features of today's SoCs are encapsulated in high-level library interfaces written by embedded-system programmers with detailed knowledge of the underlying SoC architecture. However, each generation of SoC requires a new software engineering effort due to the development of new accelerators, local and shared memories, and communication interfaces. As we build more accelerators in future SoCs, we need to answer *what kinds of programming interfaces can we give to programmers* and *what architectural improvements can make programming easier.* Therefore, there needs to be a conjoined effort at both the hardware and software layers to identify what information should be communicated across layers and how we can design hardware better to improve its programmability.

Increased architectural specialization is a likely outcome of the slowing of technology scaling in order to continue performance and energy scaling. Today's mobile SoCs provide a starting point for thinking about future architectures, but many challenges must be addressed to make specialization more pervasive and cost-effective. New architectures, design tools, and programming paradigms will be required to make this approach pervasive. This book lays out some initial research directions in these areas.

Bibliography

[1] Altera SDK for OpenCL. http://www.altera.com/products/software/opencl/opencl-index.html. 29

[2] Are 28nm Transistors the Cheapest...Forever? https://www.semiwiki.com/forum/content/2768-28nm-transistors-cheapest-forever.html. 9

[3] AxBench: Approximate Computing Benchmark. http://axbench.org/. 67

[4] Coherent Accelerator Processor Interface (CAPI) for POWER8 Systems. IBM White Paper, September 2014. 14

[5] GCC4CLI. http://gcc.gnu.org/projects/cli.html. 56

[6] Intel Historical Development Cadence. http://www.anandtech.com/show/9447/intel-10nm-and-kaby-lake. 8

[7] Intel's CEO Brian Krzanich on Q2 2015 Earnings Call. http://seekingalpha.com/article/3329035-intels-intc-ceo-brian-krzanich-on-q2-2015-results-earnings-call-transcript. 7, 8

[8] LLVM assembly language reference manual, bitcode documentation. http://llvm.org/docs/LangRef.html. 55

[9] MathWorks HDL Coder. http://www.mathworks.com/products/hdl-coder/. 29

[10] Mckinsey on Semiconductors: Moore's law: Repeal or Renewal? 8

[11] Oracle's SPARC T4 Server Architecture. Oracle White Paper, June 2012. 14, 18, 19

[12] RISC-V Rocket Core. https://github.com/ucb-bar/rocket. 31

[13] Scala Programming Language. http://www.scala-lang.org/. 30

[14] SPEC CPU2000. https://www.spec.org/cpu2000/. 67, 68

[15] SPEC CPU2006. https://www.spec.org/cpu2006/. 67, 68

[16] TI OMAP Applications Processors. http://www.ti.com/product/omap5432. 14

[17] Xilinx Vivado High-Level Synthesis. http://www.xilinx.com/products/design-tools/vivado/. 29, 33

[18] Intel 64 and ia-32 architectures software developer's manual. 2015. 14, 18, 19

[19] G. M. Amdahl. Validity of the single processor approach to achieving large scale computing capabilities. In *Proceedings of the April 18-20, 1967, Spring Joint Computer Conference*, pages 483–485. ACM, 1967. DOI: 10.1145/1465482.1465560. 6, 7

[20] W. Arden, M. Brillouët, P. Cogez, M. Graef, B. Huizing, and R. Mahnkopf. More than moore white paper. *Version*, 2:14, 2010. 9

[21] K. Asanovic, R. Bodik, B. C. Catanzaro, J. J. Gebis, P. Husbands, K. Keutzer, D. A. Patterson, W. L. Plishker, J. Shalf, S. W. Williams, and K. A. Yelick. The landscape of parallel computing research: A view from berkeley. Technical report, EECS Department, University of California, Berkeley, 2006. 69

[22] J. Auerbach, D. F. Bacon, I. Burcea, P. Cheng, S. J. Fink, R. Rabbah, and S. Shukla. A compiler and runtime for heterogeneous computing. In *Proceedings of the 49th Annual Design Automation Conference*, 2012. DOI: 10.1145/2228360.2228411. 29, 30

[23] J. Auerbach, D. F. Bacon, P. Cheng, and R. Rabbah. Lime: A java-compatible and synthesizable language for heterogeneous architectures. In *Proceedings of the ACM International Conference on Object Oriented Programming Systems Languages and Applications*. ACM, 2010. DOI: 10.1145/1869459.1869469. 29, 30

[24] T. M. Austin, E. Larson, and D. Ernst. Simplescalar: An infrastructure for computer system modeling. *IEEE Computer*, 2002. DOI: 10.1109/2.982917. 12, 33

[25] T. M. Austin and G. S. Sohi. Dynamic dependency analysis of ordinary programs. In *ISCA*, 1992. DOI: 10.1145/146628.140395. 35

[26] J. Bachrach, H. Vo, B. Richards, Y. Lee, A. Waterman, R. Avizienis, J. Wawrzynek, and K. Asanovic. Chisel: Constructing hardware in a scala embedded language. In *Design Automation Conference (DAC), 2012 49th ACM/EDAC/IEEE*, 2012. DOI: 10.1145/2228360.2228584. 29, 31

[27] A. Bakhoda, G. L. Yuan, W. W. L. Fung, H. Wong, and T. M. Aamodt. Analyzing cuda workloads using a detailed gpu simulator. In *ISPASS*, 2009. DOI: 10.1109/ISPASS.2009.4919648. 33

[28] C. Bienia, S. Kumar, J. P. Singh, and K. Li. The parsec benchmark suite: Characterization and architectural implications. In *Proceedings of the 17th International Conference on Parallel Architectures and Compilation Techniques*, 2008. DOI: 10.1145/1454115.1454128. 67

[29] N. L. Binkert, B. M. Beckmann, G. Black, S. K. Reinhardt, A. G. Saidi, A. Basu, J. Hestness, D. Hower, T. Krishna, S. Sardashti, R. Sen, K. Sewell, M. Shoaib, N. Vaish, M. D.

Hill, and D. A. Wood. The gem5 simulator. *SIGARCH Computer Architecture News*, 2011. DOI: 10.1145/2024716.2024718. 12, 33, 35

[30] B. Blaner, B. Abali, B. Bass, S. Chari, R. Kalla, S. Kunkel, K. Lauricella, R. Leavens, J. Reilly, and P. Sandon. IBM POWER7+ processor on-chip accelerators for cryptography and active memory expansion. *IBM Journal of Research and Development*, 2013. DOI: 10.1147/JRD.2013.2280090. 14

[31] M. Bohr. A 30 year retrospective on dennard's mosfet scaling paper. *Solid-State Circuits Society Newsletter, IEEE*, 2007. DOI: 10.1109/N-SSC.2007.4785534. 5

[32] D. Brooks, V. Tiwari, and M. Martonosi. Wattch: a framework for architectural-level power analysis and optimizations. In *ISCA*, 2000. DOI: 10.1145/342001.339657. 12, 33

[33] D. Burger, J. R. Goodman, and A. Kagi. Memory bandwidth limitations of future microprocessors. In *ISCA*, 1996. DOI: 10.1145/232974.232983. 47

[34] S. Campanoni, G. Agosta, S. Crespi-Reghizzi, and A. D. Biagio. A highly flexible, parallel virtual machine: Design and experience of ildjit. *Software Practice Expererience*, 2010. DOI: 10.1002/spe.950. 55

[35] J. Casper and K. Olukotun. Hardware acceleration of database operations. In *FPGA*, 2014. DOI: 10.1145/2554688.2554787. 14, 22

[36] S. Che, M. Boyer, J. Meng, D. Tarjan, J. W. Sheaffer, S.-H. Lee, and K. Skadron. Rodinia: A benchmark suite for heterogeneous computing. In *IISWC*, 2009. DOI: 10.1109/IISWC.2009.5306797. 67, 68

[37] T. Chen, Z. Du, N. Sun, J. Wang, C. Wu, Y. Chen, and O. Temam. Diannao: A small-footprint high-throughput accelerator for ubiquitous machine-learning. In *ASPLOS*, 2014. DOI: 10.1145/2541940.2541967. 14, 25

[38] Y. Chen, T. Luo, S. Liu, S. Zhang, L. He, J. Wang, L. Li, T. Chen, Z. Xu, N. Sun, and O. Teman. Dadiannao: A machine-learning supercomputer. In *MICRO*, 2014. DOI: 10.1109/MICRO.2014.58. 14, 25

[39] A. A. Chien. 10x10 must replace 90/10. In *Proceedings of the Salishan Conference on High Performance Computing*, 2010. 17

[40] A. A. Chien, D. Vasudevan, T. T. Hoang, Y. Fang, and A. Shambayati. 10x10: A case study in federated heterogeneous architecture. *IEEE Micro*, 2015. 14, 17, 18

[41] E. S. Chung, J. D. Davis, and J. Lee. Linqits: big data on little clients. *ISCA*, 2013. DOI: 10.1145/2485922.2485945. 14, 23

[42] J. Cong, Y. Fan, G. Han, and Z. Zhang. Application-specific instruction generation for configurable processor architectures. In *FPGA*, 2004. DOI: 10.1145/968280.968307. 36

[43] J. Cong, M. Ghodrat, M. Gill, B. Grigorian, H. Huang, and G. Reinman. Composable accelerator-rich microprocessor enhanced for adaptivity and longevity. In *ISLPED*, 2013. DOI: 10.1109/ISLPED.2013.6629314. 14, 20

[44] J. Cong, M. A. Ghodrat, M. Gill, B. Grigorian, and G. Reinman. Charm: a composable heterogeneous accelerator-rich microprocessor. In *ISLPED*, 2012. DOI: 10.1145/2333660.2333747. 14, 20

[45] J. Cong, K. Guruaj, M. Huang, S. Li, B. Xiao, and Y. Zou. Domain-specific processor with 3d integration for medical image processing. In *Application-Specific Systems, Architectures and Processors (ASAP), 2011 IEEE International Conference on*, 2011. DOI: 10.1109/ASAP.2011.6043279. 67

[46] P. D'Alberto, P. A. Milder, A. Sandryhaila, F. Franchetti, J. C. Hoe, J. M. F. Moura, M. Püschel, and J. Johnson. Generating fpga accelerated DFT libraries. In *IEEE Symposium on Field-Programmable Custom Computing Machines (FCCM)*, 2007. DOI: 10.1109/FCCM.2007.58. 29, 31

[47] A. Danalis, G. Marin, C. McCurdy, J. S. Meredith, P. C. Roth, K. Spafford, V. Tipparaju, and J. S. Vetter. The scalable heterogeneous computing (shoc) benchmark suite. In *Proceedings of the 3rd Workshop on General-Purpose Computation on Graphics Processing Units*, 2010. DOI: 10.1145/1735688.1735702. 34, 67, 68

[48] A. Danowitz, K. Kelley, J. Mao, J. P. Stevenson, and M. Horowitz. Cpu db: recording microprocessor history. *Communications of the ACM*, 2012. DOI: 10.1145/2133806.2133822. 6, 10

[49] R. H. Dennard, F. H. Gaensslen, H.-N. Yu, V. L. Rideout, E. Bassous, and A. R. LeBlanc. Design of ion-implanted MOSFET's with very small physical dimensions. *IEEE Journal of Solid-State Circuits*, 1974. 3

[50] L. Eeckhout, H. Vandierendonck, and K. D. Bosschere. Quantifying the impact of input data sets on program behavior and its applications. *Journal of Instruction-Level Parallelism*, 2003. 12

[51] H. Esmaeilzadeh, E. Blem, R. St. Amant, K. Sankaralingam, and D. Burger. Dark silicon and the end of multicore scaling. In *Proceedings of the 38th Annual International Symposium on Computer Architecture*, 2011. DOI: 10.1145/2024723.2000108. 6, 7, 14

[52] H. Esmaeilzadeh, A. Sampson, L. Ceze, and D. Burger. Neural acceleration for general-purpose approximate programs. In *MICRO*, 2012. DOI: 10.1109/MICRO.2012.48. 14, 17, 21, 44

[53] J. Ferrante, K. J. Ottenstein, and J. D. Warren. The program dependence graph and its use in optimization. In *Symposium on Programming*, 1984. DOI: 10.1007/3-540-12925-1_33. 36

[54] B. A. Fields, R. Bodík, and M. D. Hill. Slack: Maximizing performance under technological constraints. In *ISCA*, 2002. DOI: 10.1145/545214.545222. 35

[55] M. Fingeroff. *High-Level Synthesis Blue Book*. 2010. 36

[56] S. Garcia, D. Jeon, C. M. Louie, and M. B. Taylor. Kremlin: rethinking and rebooting gprof for the multicore age. In *PLDI*, 2011. DOI: 10.1145/1993316.1993553. 35, 39

[57] N. George, H. Lee, D. Novo, T. Rompf, K. J. Brown, A. K. Sujeeth, M. Odersky, K. Olukotun, and P. Ienne. Hardware system synthesis from domain-specific languages. In *Field Programmable Logic and Applications (FPL), 2014 24th International Conference on*, 2014. DOI: 10.1109/FPL.2014.6927454. 29, 30

[58] V. Govindaraju, C.-H. Ho, and K. Sankaralingam. Dynamically specialized datapaths for energy efficient computing. In *HPCA*, 2011. DOI: 10.1109/HPCA.2011.5749755. 14, 16, 68

[59] V. Govindaraju, T. Nowatzki, and K. Sankaralingam. Breaking simd shackles with an exposed flexible microarchitecture and the access execute pdg. In *PACT*, 2013. DOI: 10.1109/PACT.2013.6618830. 36

[60] M. Guthaus, J. Ringenberg, D. Ernst, T. Austin, T. Mudge, and R. Brown. Mibench: A free, commercially representative embedded benchmark suite. In *Workload Characterization, 2001. WWC-4. 2001 IEEE International Workshop on*, 2001. DOI: 10.1109/WWC.2001.15. 67

[61] R. Hameed, W. Qadeer, M. Wachs, O. Azizi, A. Solomatnikov, B. C. Lee, S. Richardson, C. Kozyrakis, and M. Horowitz. Understanding sources of inefficiency in general-purpose chips. In *ISCA*, 2010. DOI: 10.1145/1816038.1815968. 14, 18

[62] Y. Hara, H. Tomiyama, S. Honda, H. Takada, and K. Ishii. Chstone: A benchmark program suite for practical c-based high-level synthesis. In *ISCAS*, 2008. DOI: 10.1109/ISCAS.2008.4541637. 67

[63] J. Hegarty, J. Brunhaver, Z. DeVito, J. Ragan-Kelley, N. Cohen, S. Bell, A. Vasilyev, M. Horowitz, and P. Hanrahan. Darkroom: Compiling high-level image processing code into hardware pipelines. In *SIGGRAPH*, 2014. DOI: 10.1145/2601097.2601174. 29

[64] K. Hoste and L. Eeckhout. Comparing benchmarks using key microarchitecture-independent characteristics. In *International Symposium on Workload Characterization*, 2006. DOI: 10.1109/IISWC.2006.302732. 12, 51

[65] W. Hunt, B. A. Maher, D. Burger, and K. S. Mckinley. Optimal huffman tree-height reduction for instruction-level parallelism. *Technical Report TR-08-34, Department of Computer Sciences The University of Texas at Austin*, 2008. 37

[66] H. C. Hunter and W. mei W. Hwu. Code coverage and input variability: effects on architecture and compiler research. In *CASES*, 2002. DOI: 10.1145/581630.581643. 43

[67] T. Inc. How to minimize energy consumption while maximizing asic and soc performance. http://ip.cadence.com/uploads/white_papers/Xenergy_Tensilica.pdf 17

[68] D. Jeon, S. Garcia, C. M. Louie, and M. B. Taylor. Kismet: parallel speedup estimates for serial programs. In *OOPSLA*, 2011. DOI: 10.1145/2048066.2048108. 35

[69] R. Kessler. The Cavium 32 Core OCTEON II 68xx. *Hop Chips*, 2011. 14, 18, 19

[70] M. Kim, H. Kim, and C.-K. Luk. Sd3: A scalable approach to dynamic data-dependence profiling. In *MICRO*, 2010. DOI: 10.1109/MICRO.2010.49. 39

[71] A. Krishna, T. Heil, N. Lindberg, F. Toussi, and S. VanderWiel. Hardware acceleration in the ibm poweren processor: Architecture and performance. In *Proceedings of the 21st International Conference on Parallel Architectures and Compilation Techniques*, 2012. DOI: 10.1145/2370816.2370872. 14

[72] M. Kumar. Measuring parallelism in computation-intensive scientific/engineering applications. *IEEE Trans. Computers*, 1988. DOI: 10.1109/12.2259. 36

[73] M. S. Lam and R. P. Wilson. Limits of control flow on parallelism. In *ISCA*, 1992. DOI: 10.1145/146628.139702. 38

[74] C. Lee, M. Potkonjak, and W. Mangione-Smith. Mediabench: a tool for evaluating and synthesizing multimedia and communications systems. In *Microarchitecture, 1997. Proceedings., Thirtieth Annual IEEE/ACM International Symposium on*, 1997. DOI: 10.1109/MICRO.1997.645830. 67

[75] H.-H. S. Lee, M. Smelyanskiy, C. J. Newburn, and G. S. Tyson. Stack value file: custom microarchitecture for the stack. In *International Symposium on High Performance Computer Architecture*, 2001. DOI: 10.1109/HPCA.2001.903247. 53

[76] Y. Lee, A. Waterman, R. Avizienis, henry Cook, C. Sun, V. Stojanov, and K. Asanovic. A 45nm 1.3ghz 16.7 double-precision gflops/w risc-v processor with vector accelerators. In *ESSCIRC*, 2014. DOI: 10.1109/ESSCIRC.2014.6942056. 14, 20, 31

[77] J. Leng, T. H. Hetherington, A. ElTantawy, S. Z. Gilani, N. S. Kim, T. M. Aamodt, and V. J. Reddi. Gpuwattch: enabling energy optimizations in gpgpus. In *ISCA*, 2013. DOI: 10.1145/2508148.2485964. 33

[78] M.-L. Li, R. Sasanka, S. Adve, Y.-K. Chen, and E. Debes. The alpbench benchmark suite for complex multimedia applications. In *Workload Characterization Symposium, 2005. Proceedings of the IEEE International*, 2005. 67

[79] S. Li, J. H. Ahn, R. D. Strong, J. B. Brockman, D. M. Tullsen, and N. P. Jouppi. Mcpat: an integrated power, area, and timing modeling framework for multicore and manycore architectures. In *MICRO*, 2009. DOI: 10.1145/1669112.1669172. 12, 33

[80] K. T. Lim, D. Meisner, A. G. Saidi, P. Ranganathan, and T. F. Wenisch. Thin servers with smart pipes: designing soc accelerators for memcached. In *ISCA*, 2013. DOI: 10.1145/2485922.2485926. 14, 21, 22, 44

[81] H.-Y. Liu and L. P. Carloni. On learning-based methods for design-space exploration with high-level synthesis. In *DAC*, 2013. DOI: 10.1145/2463209.2488795. 46

[82] D. Lockhart, G. Zibrat, and C. Batten. Pymtl: A unified framework for vertically integrated computer architecture research. In *47th IEEE/ACM Int'l Symp. on Microarchitecture (MICRO)*, Dec 2014. DOI: 10.1109/MICRO.2014.50. 29, 31

[83] S. Lu, P. Zhou, W. Liu, Y. Zhou, and J. Torrellas. Pathexpander: Architectural support for increasing the path coverage of dynamic bug detection. In *MICRO*, 2006. DOI: 10.1109/MICRO.2006.40. 43

[84] C.-K. Luk, R. Cohn, R. Muth, H. Patil, A. Klauser, G. Lowney, S. Wallace, V. J. Reddi, and K. Hazelwood. Pin: building customized program analysis tools with dynamic instrumentation. *PLDI*, 2005. DOI: 10.1145/1065010.1065034. 49

[85] C.-K. Luk, R. Cohn, R. Muth, H. Patil, A. Klauser, G. Lowney, S. Wallace, V. J. Reddi, and K. Hazelwood. Pin: Building customized program analysis tools with dynamic instrumentation. In *PLDI*, 2005. DOI: 10.1145/1065010.1065034. 56

[86] M. J. Lyons, M. Hempstead, G.-Y. Wei, and D. Brooks. The accelerator store: A shared memory framework for accelerator-based systems. *TACO*, 2012. DOI: 10.1145/2086696.2086727. 14, 24

[87] H. Mao, S. Karandikar, A. Ou, and S. Basu. Hardware acceleration of key-value stores. *UC Berkeley CS262a Report*, 2014. 14, 22, 31

[88] P. A. Milder, F. Franchetti, J. C. Hoe, and M. Püschel. Computer generation of hardware for linear digital signal processing transforms. *ACM Transactions on Design Automation of Electronic Systems*, 2012. DOI: 10.1145/2159542.2159547. 29, 31

[89] G. E. Moore. Cramming more components onto integrated circuits. *Electronics Magazine*, 1965. DOI: 10.1109/JPROC.1998.658762. 2

[90] G. E. Moore. No exponential is forever: but "forever" can be delayed! In *IEEE International Solid-State Circuits Conference*, 2003. DOI: 10.1109/ISSCC.2003.1234194. 5

[91] T. Moreau, M. Wyse, jacob Nelson, A. Sampson, H. Esmaeilzadeh, L. Ceze, and M. Oskin. SNNAP: Approximate computing on programmable socs via neural acceleration. In *HPCA*, 2015. DOI: 10.1109/JPROC.1998.658762. 14, 21

[92] J. M. F. Moura, J. Johnson, R. W. Johnson, D. Padua, V. K. Prasanna, M. Püschel, B. Singer, M. Veloso, and J. Xiong. Generating platform-adapted DSP libraries using SPIRAL. In *High Performance Embedded Computing (HPEC)*, 2001. 31

[93] R. Narayanan, B. Ozisikyilmaz, J. Zambreno, G. Memik, and A. Choudhary. Minebench: A benchmark suite for data mining workloads. In *2006 IEEE International Symposium on Workload Characterization*, 2006. DOI: 10.1109/IISWC.2006.302743. 67, 68

[94] R. S. Nikhil. Abstraction in hardware system design. *ACM Queue*, 2011. DOI: 10.1145/2016036.2020861. 29

[95] A. Ou, Q. Nguyen, Y. Lee, and K. Asanovic. A case for mvps: Mixed-precision vector processors. In *ISCA Parallelism in Mobile Platforms Workshop*, 2014. 14, 19, 20

[96] A. Papakonstantinou, K. Gururaj, J. Stratton, D. Chen, J. Cong, and W.-M. Hwu. Fcuda: Enabling efficient compilation of cuda kernels onto fpgas. In *Application Specific Processors, 2009. SASP '09. IEEE 7th Symposium on*, 2009. DOI: 10.1109/SASP.2009.5226333. 29

[97] A. Putnam, A. Caulfield, E. Chung, D. Chiou, K. Constantinides, J. Demme, H. Esmaeilzadeh, J. Fowers, G. Gopal, J. Gray, M. Haselman, S. Hauck, S. Heil, A. Hormati, J.-Y. Kim, S. Lanka, J. Larus, E. Peterson, S. Pope, A. Smith, J. Thong, P. Xiao, and D. Burger. A reconfigurable fabric for accelerating large-scale datacenter services. In *Computer Architecture (ISCA), 2014 ACM/IEEE 41st International Symposium on*, 2014. DOI: 10.1109/ISCA.2014.6853195. 14, 26

[98] W. Qadeer, R. Hameed, O. Shacham, P. Venkatesan, C. Kozyrakis, and M. A. Horowitz. Convolution engine: balancing efficiency & flexibility in specialized computing. In *ISCA*, 2013. DOI: 10.1145/2735841. 14, 17, 18

[99] L. Rauchwerger, P. K. Dubey, and R. Nair. Measuring limits of parallelism and characterizing its vulnerability to resource constraints. In *MICRO*, 1993. DOI: 10.1145/255235.255268. 35

[100] B. Reagen, R. Adolf, Y. Shao, G.-Y. Wei, and D. Brooks. Machsuite: Benchmarks for accelerator design and customized architectures. In *Workload Characterization (IISWC), 2014 IEEE International Symposium on*, 2014. DOI: 10.1109/IISWC.2014.6983050. 67, 68, 69

[101] P. Rosenfeld, E. Cooper-Balis, and B. Jacob. Dramsim2: A cycle accurate memory system simulator. *IEEE Computer Architecture Letters*, 2011. DOI: 10.1109/L-CA.2011.4. 34, 46

[102] J. Sampson, G. Venkatesh, N. Goulding-Hotta, S. Garcia, S. Swanson, and M. B. Taylor. Efficient complex operators for irregular codes. In *HPCA*, 2011. DOI: 10.1109/HPCA.2011.5749754. 20

[103] R. Sampson, M. Yang, S. Wei, C. Chakrabarti, and T. F. Wenisch. Sonic millip3de: A massively parallel 3d-stacked accelerator for 3d ultrasound. In *HPCA*, 2013. DOI: 10.1109/HPCA.2013.6522329. 14, 24, 68

[104] B. Schaller. The origin, nature, and implications of moore's law. 2014. http://research.microsoft.com/en-us/um/people/gray/Moore_Law.html 3

[105] T. Sha, M. M. K. Martin, and A. Roth. Nosq: Store-load communication without a store queue. In *In MICRO*, pages 285–296, 2006. DOI: 10.1109/MICRO.2006.39. 38

[106] O. Shacham, M. Wachs, A. Danowitz, S. Galal, J. Brunhaver, W. Qadeer, S. Sankaranarayanan, A. Vassiliev, S. Richardson, and M. Horowitz. Avoiding game over: Bringing design to the next level. In *Proceedings of the 49th Annual Design Automation Conference*, 2012. DOI: 10.1145/2228360.2228472. 29

[107] C. Shannon. A mathematical theory of communication. *Bell System Technical Journal*, 1948. DOI: 10.1002/j.1538-7305.1948.tb00917.x. 61

[108] Y. S. Shao and D. Brooks. Isa-independent workload characterization and its implications for specialized architectures. In *ISPASS*, 2013. DOI: 10.1109/ISPASS.2013.6557175. 36

[109] Y. S. Shao, B. Reagen, G.-Y. Wei, and D. Brooks. Aladdin: A pre-rtl, power-performance accelerator simulator enabling large design space exploration of customized architectures. In *ISCA*, 2014. DOI: 10.1109/ISCA.2014.6853196. 30, 34, 35, 40, 42, 44, 45, 47, 48, 49

[110] T. Sherwood, E. Perelman, G. Hamerly, and B. Calder. Automatically characterizing large scale program behavior. In *ASPLOS*, 2002. DOI: 10.1145/605397.605403. 56

[111] A. Solomatnikov, A. Firoozshahian, O. Shacham, Z. Asgar, M. Wachs, W. Qadeer, S. Richardson, and M. Horowitz. Using a configurable processor generator for computer architecture prototyping. In *MICRO*, 2009. DOI: 10.1145/1669112.1669159. 19

[112] J. A. Stratton, C. Rodrigrues, I.-J. Sung, N. Obeid, L. Chang, G. Liu, and W.-M. W. Hwu. Parboil: A revised benchmark suite for scientific and commercial throughput computing. University of Illinois at Urbana-Champaign, 2012. 12, 67

[113] A. Sujeeth, H. Lee, K. Brown, T. Rompf, H. Chafi, M. Wu, A. Atreya, M. Odersky, and K. Olukotun. Optiml: an implicitly parallel domain-specific language for machine learning. In *Proceedings of the 28th International Conference on Machine Learning (ICML-11)*, 2011. 29, 30

[114] A. K. Sujeeth, K. J. Brown, H. Lee, T. Rompf, H. Chafi, M. Odersky, and K. Olukotun. Delite: A compiler architecture for performance-oriented embedded domain-specific languages. *ACM Transactions on Embedded Computing Systems (TECS)*, 2014. DOI: 10.1145/2584665. 29, 30

[115] K. B. Theobald, G. R. Gao, and L. J. Hendren. On the limits of program parallelism and its smoothability. In *MICRO*, 1992. DOI: 10.1145/144965.144977. 37

[116] S. Thomas, C. Gohkale, E. Tanuwidjaja, T. Chong, D. Lau, S. Garcia, and M. Bedford Taylor. Cortexsuite: A synthetic brain benchmark suite. In *Workload Characterization (IISWC), 2014 IEEE International Symposium on*, 2014. DOI: 10.1109/IISWC.2014.6983043. 67

[117] M. A. Unger, H.-P. Chou, T. Thorsen, A. Scherer, and S. R. Quake. Monolithic microfabricated valves and pumps by multilayer soft lithography. *Science*, 288(5463):113–116, 2000. DOI: 10.1126/science.288.5463.113. 8

[118] S. K. Venkata, I. Ahn, D. Jeon, A. Gupta, C. Louie, S. Garcia, S. Belongie, and M. B. Taylor. Sd-vbs: The san diego vision benchmark suite. *IISWC*, 2009. 12, 67, 68

[119] G. Venkatesh, J. Sampson, N. Goulding, S. Garcia, V. Bryksin, J. Lugo-Martinez, S. Swanson, and M. B. Taylor. Conservation cores: reducing the energy of mature computations. *ASPLOS*, 2010. DOI: 10.1145/1735970.1736044. 14, 20, 21, 57

[120] G. Venkatesh, J. Sampson, N. Goulding-Hotta, S. K. Venkata, M. B. Taylor, and S. Swanson. Qscores: trading dark silicon for scalable energy efficiency with quasi-specific cores. In *MICRO*, 2011. DOI: 10.1145/2155620.2155640. 20

[121] H. Vo, Y. Lee, A. Waterman, and krste Asanovic. A case for os-friendly hardware accelerators. In *ISCA Interaction Between Operating System and Computer Architecture Workshop*, 2013. 14, 20

[122] D. W. Wall. Limits of instruction-level parallelism. In *ASPLOS*, 1991. DOI: 10.1145/106972.106991. 35, 36

[123] S. J. E. Wilton and N. P. Jouppi. Cacti: An enhanced cache access and cycle time model. *IEEE Journal of Solid-State Circuits*, 1996. DOI: 10.1109/4.509850. 43

[124] S. C. Woo, M. Ohara, E. Torrie, J. P. Singh, and A. Gupta. The splash-2 programs: Characterization and methodological considerations. In *Proceedings of the 22nd Annual International Symposium on Computer Architecture*, 1995. DOI: 10.1109/ISCA.1995.524546. 67

[125] L. Wu, R. J. Barker, M. A. Kim, and K. A. Ross. Navigating big data with high-throughput, energy-efficient data partitioning. In *ISCA*, 2013. DOI: 10.1145/2485922.2485944. 14, 25, 44, 68

[126] L. Wu, A. Lottarini, T. K. Paine, M. A. Kim, and K. A. Ross. Q100: The architecture and design of a database processing unit. In *ASPLOS*, 2014. DOI: 10.1145/2541940.2541961. 14, 23

[127] L. Yen, S. C. Draper, and M. D. Hill. Notary: Hardware techniques to enhance signatures. In *International Symposium on Microarchitecture*, 2008. DOI: 10.1109/MICRO.2008.4771794. 62

[128] T. Yokota, K. Ootsu, and T. Baba. Introducing entropies for representing program behavior and branch predictor performance. In *Workshop on Experimental Computer Science*, 2007. DOI: 10.1145/1281700.1281717. 64

[129] K. Yuan, J.-S. Yang, and D. Z. Pan. Double patterning layout decomposition for simultaneous conflict and stitch minimization. *Computer-Aided Design of Integrated Circuits and Systems, IEEE Transactions on*, 29(2):185–196, 2010. DOI: 10.1109/TCAD.2009.2035577. 8

[130] N. Zhang and B. Brodersen. The cost of flexibility in systems on a chip design for signal processing applications. *University of California, Berkeley, Tech. Rep*, 2002. 11

[131] Q. Zhu, B. Akin, H. E. Sumbul, F. Sadi, J. Hoe, L. Pileggi, and F. Franchetti. A 3d-stacked logic-in-memory accelerator for application-specific data intensive computing. In *Proceedings of IEEE International 3D Systems Integration Conference (3DIC)*, 2013. DOI: 10.1109/3DIC.2013.6702348. 38

Authors' Biographies

YAKUN SOPHIA SHAO

Yakun Sophia Shao is a Ph.D. candidate in Computer Science at Harvard University, working with Professor David Brooks. Her primary research interests lie in the general area of computer architecture, with a particular focus on modeling and design for heterogeneous architectures. She received her B.S. degree in Electrical Engineering from Zhejiang University, China and her S.M. degree in Computer Science from Harvard University. Her paper was selected as one of the Top Picks in Computer Architecture in 2014. She is a Siebel Scholar and a recipient of the IBM Ph.D. Fellowship.

DAVID BROOKS

David Brooks is the Haley Family Professor of Computer Science in the School of Engineering and Applied Sciences at Harvard University. He joined Harvard in 2002 after spending one year as a research staff member at IBM T.J. Watson Research Center. Professor Brooks received his B.S. in Electrical Engineering at the University of Southern California and M.A. and Ph.D. degrees in Electrical Engineering at Princeton University.

Professor Brooks has received several honors and awards including the ACM Maurice Wilkes Award, NSF CAREER award, IBM Faculty Partnership Award, and DARPA Young Faculty Award. He has received best paper awards at MICRO, HPCA, and ICCD and has had several papers selected for IEEE Micro's "Top Picks in Computer Architecture" since 2005. His research interests include technology-aware computer design, with an emphasis on power-efficient computer architectures for high-performance and embedded systems.

Printed in the United States
by Baker & Taylor Publisher Services